Measuring the Science and Engineering Enterprise

PRIORITIES FOR THE DIVISION OF SCIENCE RESOURCES STUDIES

Committee to Assess the Portfolio of
The Division of Science Resources Studies of NSF

Office of Scientific and Engineering Personnel
Committee on National Statistics

National Research Council

NATIONAL ACADEMY PRESS
Washington, D.C.

NOTICE: The project that is the subject of this report was approved by the Governing Board of the National Research Council, whose members are drawn from the councils of the National Academy of Sciences, the National Academy of Engineering, and the Institute of Medicine. The members of the committee responsible for the report were chosen for their special competencies and with regard for appropriate balance.

This report has been reviewed by a group other than the authors according to procedures approved by a Report Review Committee consisting of members of the National Academy of Sciences, the National Academy of Engineering, and the Institute of Medicine.

This study was supported by Contract No. SRS-9802651 between the National Academy of Sciences and the National Science Foundation. Any opinions, findings, conclusions, or recommendations expressed in this publication are those of the author(s) and do not necessarily reflect the views of the organizations or agencies that provided support for the project.

International Standard Book Number 0-309-06892-4

Additional copies of this report are available:
National Academy Press (http://www.nap.edu)
2101 Constitution Avenue, N.W., Box 285
Washington, D.C. 20055
800-624-6242
202-334-3313 (in the Washington metropolitan area)

THE NATIONAL ACADEMIES

National Academy of Sciences
National Academy of Engineering
Institute of Medicine
National Research Council

The **National Academy of Sciences** is a private, nonprofit, self-perpetuating society of distinguished scholars engaged in scientific and engineering research, dedicated to the furtherance of science and technology and to their use for the general welfare. Upon the authority of the charter granted to it by the Congress in 1863, the Academy has a mandate that requires it to advise the federal government on scientific and technical matters. Dr. Bruce M. Alberts is president of the National Academy of Sciences.

The **National Academy of Engineering** was established in 1964, under the charter of the National Academy of Sciences, as a parallel organization of outstanding engineers. It is autonomous in its administration and in the selection of its members, sharing with the National Academy of Sciences the responsibility for advising the federal government. The National Academy of Engineering also sponsors engineering programs aimed at meeting national needs, encourages education and research, and recognizes the superior achievements of engineers. Dr. William A. Wulf is president of the National Academy of Engineering.

The **Institute of Medicine** was established in 1970 by the National Academy of Sciences to secure the services of eminent members of appropriate professions in the examination of policy matters pertaining to the health of the public. The Institute acts under the responsibility given to the National Academy of Sciences by its congressional charter to be an adviser to the federal government and, upon its own initiative, to identify issues of medical care, research, and education. Dr. Kenneth I. Shine is president of the Institute of Medicine.

The **National Research Council** was organized by the National Academy of Sciences in 1916 to associate the broad community of science and technology with the Academy's purposes of furthering knowledge and advising the federal government. Functioning in accordance with general policies determined by the Academy, the Council has become the principal operating agency of both the National Academy of Sciences and the National Academy of Engineering in providing services to the government, the public, and the scientific and engineering communities. The Council is administered jointly by both Academies and the Institute of Medicine. Dr. Bruce M. Alberts and Dr. William A. Wulf are chairman and vice chairman, respectively, of the National Research Council.

Office of Scientific and Engineering Personnel
1998-1999 Advisory Committee

M. R. C. Greenwood, University of California, Santa Cruz (*Chair*)
John D. Wiley, University of Wisconsin, Madison (*Vice Chair*)
Ronald G. Ehrenberg, Cornell University
Carlos Gutierrez, California State University
Stephen J. Lukasik, Independent Consultant
William H. Miller, University of California, Berkeley (ex officio)
Claudia Mitchell-Kernan, University of California, Los Angeles
Edward Penhoet, University of California, Berkeley
Tadataka Yamada, SmithKline Beecham Corporation
A. Thomas Young, Lockheed Martin Corporation (retired)

Charlotte Kuh, *Executive Director*
Marilyn Baker, *Associate Executive Director*

Committee on National Statistics
1998-1999

John E. Rolph, University of Southern California (*Chair*)
Joseph Altonji, Northwestern University.
Julie Davanzo, The RAND Corporation
William Eddy, Carnegie Mellon University
William D. Kalsbeek, University of North Carolina-Chapel Hill
Roderick Little, University of Michigan
Thomas A. Louis, University of Minnesota
Charles F. Manski, Northwestern University
William Nordhaus, Yale University
Janet Norwood, the Urban Institute
Edward B. Perrin, University of Washington
Paul Rosenbaum, University of Pennsylvania
Francisco Samaniego, University of California, Davis
Richard Schmalensee, Massachusetts Institute of Technology

Andrew A. White, *Acting Director*

Acknowledgements

This report has benefited from input from various individuals. The committee thanks Bennett I. Bertenthal, former NSF Assistant Director for the Directorate of Social, Behavioral, and Economic Sciences, for sharing his insights with the committee and providing support for this study. The committee acknowledges division and program managers in the Division of Science Resources Studies who made presentations to the committee: Jeanne E. Griffith, former Director; Alan R. Tupek, former Deputy Director; Ronald S. Fecso, Chief Mathematical Statistician; Jennifer S. Bond, Director, Science and Engineering Indicators Program; Mary J. Golladay, Director, Human Resources Statistics Program; John E. Jankowski, Director, Research and Development Statistics Program; and Rolf F. Lehming, Director, Integrated Studies Program. The committee also acknowledges SRS staff who participated in focus groups for this study: Carolyn B. Arena, Richard J. Bennof, Joan S. Burrelli, Mary V. Burke, Lawrence Burton, Joanne P. Carr, Deborah A. Collins, Eileen Collins, John R. Gawalt, Tanya R. Gore, Julia Harriston, Jennifer R. Held, Susan T. Hill, Anne M. Houghton, Theodosia L. Jacobs, Jean M. Johnson, Kelly H. Kang, Ann T. Lanier, Mary M. Machen, Ronald L. Meeks, Richard E. Morrison, Melissa F. Pollak, Alan I. Rapoport, Lawrence M. Rausch, Mark C. Regets, Steven Payson, R. Keith Wilkinson, and Raymond M. Wolfe.

The committee also acknowledges those who made presentations at the workshop or made themselves available for interviews: James Adams, University of Florida; Thomas Arrison, National Research Council; John Auerbach, Council on Competitiveness; Wendy Baldwin, National Institutes of Health; Arthur Bienenstock, White House Office of Science and Technology Policy; John Birge, University of Michigan; William Blanpied, National Science Foundation; Erich Bloch, Council on Competitiveness; William Boesman, Congressional Research Service; Joseph Bordogna, National Science Foundation; John Boright, National Research Council; Lewis Branscomb, Harvard University; Marta Cehelsky, National Science Board; Jasemine Chambers, U.S. Patent and Trademark Office; Elinor Champion, U.S. Bureau of the Census; Joseph Clark, American Chemical Society; Mary Clutter, National Science Foundation; Linda Cohen, University of California, Irvine; Robert W. Correll, National Science Foundation; Brenda Cox, Mathematica Policy Research; Robert Dauffenbach, University of Oklahoma; Denice Denton, University of Washington; Paul Doremus, National Institute of Standards and Technology; Steve Eule, Committee on Science, U.S. House of Representatives; Irwin Feller, Pennsylvania State University; Michael Finn, Oak Ridge Institute for Science and Education; Clifford Gabriel, White House Office of Science and Technology Policy; Catherine Gaddy, Commission on Professionals in Science and Technology; Howard Garrison, Federation of American Societies for Experimental Biology; Fred Gault, Statistics Canada; Gerald Hane, White House Office of Science and Technology Policy; Sarah Horrigan, U.S. Office of Management and Budget; Susanne Huttner, University of California System; Adam Jaffe, Brandeis University; Ronald Jarmin, Center for Economic Studies, U.S. Bureau of the Census; Joseph Jasinski, IBM Research; Gretchen Jordan, Sandia National Laboratories; Thomas Kalil, National Economic Council, Executive Office of the President; Philip Kiko, Committee on Science, U.S. House of Representatives; Kei Koizumi, American Association for the Advancement of Science; Stephen J. Lukasik, Independent Consultant; Pete Maggiore, State of New Mexico; Shirley Malcom, American Association for the Advancement of Science; Daniel Malkin, Organization for Economic Cooperation and Development; Dominique Martin-Rovet, Embassy of France; Bernard McDonald, National Science Foundation; Ray Merenstein, Research!America; Susan Mitchell, Mathematica Policy Research; Duncan Moore, White House Office of Science and Technology Policy; Julia Moore, National Science Foundation; Stephen Nelson, American Association for the Advancement of

Science; Janet Norwood, The Urban Institute; Sumiye Okubo, Bureau of Economic Analysis, U.S. Department of Commerce; Linda Parker, National Science Foundation; David Radzanowski, U.S. Office of Management and Budget; J. Thomas Ratchford, George Mason University; Proctor Reid, National Academy of Engineering; Sherwin Rosen, University of Chicago; Michael Seiverts, National Science Foundation; Leslie B. Sims, University of Iowa; Deborah Stine, National Academy of Sciences; Michael Teitelbaum, Sloan Foundation; Charles Vest, Massachusetts Institute of Technology; James Voytuk, National Research Council; Katherine Wallman, U.S. Office of Management and Budget; Daniel Werfel, U.S. Office of Management and Budget; James D. Wilson, Committee on Science, U.S. House of Representatives; and Patrick Windham, R. Wayne Sayers and Associates. The committee also wishes to thank the Council of Graduate Schools for its help in organizing two sessions at its summer workshop for graduate deans in Santa Fe, New Mexico in 1998 where this study was also discussed.

This report has been reviewed by persons chosen for their diverse perspectives and technical expertise in accordance with procedures approved by the National Research Council's Report Review Committee. The purposes of this independent review are to provide candid and critical comments that will assist the study committee in making its report as sound as possible and to ensure that the report meets institutional standards of objectivity, evidence, and responsiveness to the study charge. The review comments and draft manuscript remain confidential to protect the integrity of the deliberative process. We wish to thank the following for their participation in the review of this report: Richard Attiyeh, University of California, San Diego; Daniel Berg, Rensselaer Polytechnic Institute; Julie DaVanzo, the RAND Corporation; William F. Eddy, Carnegie-Mellon University; Michael Finn, Oak Ridge Institute for Science and Technology; Christopher Hill, George Mason University; C. Dan Mote, University of Maryland, College Park; and the report review coordinator, A. Thomas Young. Although those persons have provided many constructive comments and suggestions, responsibility for the final content of this report rests solely with the study committee.

The project and production of the report was aided by the invaluable help of National Research Council professional staff, principally Peter Henderson, study director, Education and Career Studies Unit, Office of Scientific and Engineering Personnel, and Contance Citro, senior program officer, Committee on National Statistics. Peter devoted countless hours to organizing the wide ranging ideas of the committee into a coherent structure and to crafting prose that appropriately transmitted the nuances of complex committee discussions. His patience and energy were critical to the creation of a committee consensus. Margaret Boone of Policy Research Methods, Inc., Thomas Arrison of the NRC's Policy Division, and Dr. Citro conducted interviews and focus groups for this study. Edvin Hernandez ably organized meetings of the study committee. Martha Bohman provided administrative support for the study.

The committee also wishes to express its thanks to Charlotte Kuh, executive director of OSEP, and Marilyn Baker, associate executive director of OSEP, for their valuable insights on the issues the study addresses.

Janice Madden
Chair
Committee to Assess the Portfolio of the
Science Resources Studies Division of NSF

Contents

References

Appendices

List of Tables

List of Figures

List of Boxes

Abbreviations in the Report

AEA	American Economics Association
AAAS	American Association for the Advancement of Science
AGS	Association of Graduate Schools
AAU	Association of American Universities
BLS	U.S. Bureau of Labor Statistics
CATI	Computer-assisted telephone interviewing
CGS	Council of Graduate Schools
CNSTAT	NRC Committee on National Statistics
COSEPUP	Committee on Science, Engineering, and Public Policy
CPST	Commission on Professionals in Science and Technology
CRADA	Cooperative Research and Development Agreement
CRS	Congressional Research Service
DARPA	Defense Advanced Research Projects Agency
DOD	U.S. Department of Defense
DOE	U.S. Department of Energy
DDP	Doctorate Data Project
ETS	Educational Testing Service
FFRDCs	Federally funded research and development centers
FS&T	Federal science and technology (budget)
FTE	Full-time equivalent
GRE	Graduate Record Examination
GSPSE	Survey of Graduate Students and Postdoctorats in Science and Engineering
HRS	Human Resources Statistics
IPEDS	Integrated Postsecondary Education Data System
IT	Information Technology
NAS	National Academy of Sciences
NASA	National Aeronautics and Space Administration
NCES	National Center for Education Statistics
NCRA	National Cooperative Research Act of 1994
NEC.	Not elsewhere classified
NEH	National Endowment for the Humanities
NIH	National Institutes of Health
NRC	National Research Council
NRSEP	National Register of Scientific and Engineering Personnel
NSCG	National Survey of College Graduates
NSCRG	National Survey of Recent College Graduates
NSB	National Science Board
NSF	National Science Foundation
OMB	U.S. Office of Management and Budget
OSTP	U.S. Office of Science and Technology Policy
PI	Principal Investigator
PRA	NSF Division of Policy Research and Analysis
R&D	Research and Development
RD-1	Survey of Industrial Research and Development
RDS	Research and Development Statistics
RJV	Research Joint Venture
S&E	Science and Engineering
SBE	Directorate of Social, Behavioral, and Economic Sciences
SED	Survey of Earned Doctorates
SESTAT	Scientists and Engineers Statistical Data System
SDR	Survey of Doctorate Recipients
SRS	Division of Science Resources Studies
STEP	NRC Board on Science, Technology, and Economic Policy
STPDS	Scientific and Technical Personnel Data System
WebCASPAR	Web-based Computer-Assisted Science Policy Analysis and Research System

Executive Summary

The Division of Science Resources Studies (SRS) of the National Science Foundation (NSF) is one of fourteen agencies in the federal statistical system, as represented on the Interagency Council on Statistical Policy. SRS is charged with providing data and analyses on various policy areas for public- and private-sector constituents. As mandated by the National Science Foundation Act of 1950 as amended, SRS is "to provide a central clearinghouse for the collection, interpretation, and analysis of data on scientific and engineering resources, and to provide a source of information for policy formulation by other agencies of the federal government." To fulfill this mandate, SRS collects and acquires data on national patterns of research and development funding and performance, and on the education and careers of scientists and engineers.

To keep its data and analysis relevant to policymakers, educators, managers, and researchers, SRS should continually review and update the concepts it seeks to measure, and revise its survey instruments, survey operations, and data analysis as needed to keep them current. To achieve this, SRS must strengthen the frequency and intensity of its dialogue and interactions with its data users and develop internal processes to convert the feedback it receives from these stakeholders into changes in its surveys and analyses. As a key element of this strategy, SRS should create advisory committees for SRS surveys that would assist SRS in establishing priorities for future change. SRS should also actively engage external researchers in both the development of its surveys and the analysis of its data to broaden the range of expertise that is brought to bear on them. SRS should further increase the analytic value of its data by finding ways to improve comparability and linkages among its data sets and by continually improving the timeliness with which its data are released. All of these strategies—creating a dialogue, instituting advisory committees, engaging external researchers, linking data, and improving timeliness—should be used in addressing important, current science and engineering resources issues that are described briefly below and more fully in the report. Finally, we expect SRS to meet accepted standards for statistical agencies in independence, professional staffing, data quality, and data analysis.

Scope of Study

Today, SRS is organized functionally into data collection, analysis, and dissemination units. Two statistical programs—the Human Resources Statistics Program (HRS) and the Research and

1

Development Statistics Program (RDS)—have responsibility for primary data collection and acquisition within SRS on personnel, education, and research and development (R&D) funding and performance. HRS and RDS also produce data tabulations and reports for dissemination to the public. Two additional units—the Science and Engineering Indicators Program, responsible for supporting the National Science Boards' biennial *Science and Engineering Indicators,* and the Integrated Studies Program, responsible for producing special analyses on science and engineering resource topics—focus primarily on analysis activities. The division is supported by the Information Services Group, which oversees publication of SRS reports, both in hard copy and via the SRS web site.

To be effective in this role in the long term, SRS must ensure the ongoing relevance of the information it provides through its portfolio of data collection and analysis activities. Indeed, recent developments in the science and engineering enterprise have suggested the need for SRS to update its portfolio. Patterns in industrial R&D investment have changed. For example, industry's share of R&D funding in the United States has increased from about half in 1980 to almost two-thirds today. Service sector R&D, which accounted for just 5 percent of industrial R&D in the early 1980s, now makes up almost 25 percent of it. Federal R&D spending since the end of the Cold War has been characterized by decreases in defense R&D and an acceleration in the long-term growth of federal spending on biomedical research. Similarly, patterns in the education and careers of scientists and engineers have also changed. For example, new science Ph.D.s in some fields have encountered difficulties in the job market in the 1990s, with increased numbers of recent Ph.D.s in a series of postdoctoral positions and others working outside their fields of study. Since

the early 1990s, more than half of doctorate-level scientists and engineers work in positions outside of academia. Still other developments suggest the structure of the science and engineering enterprise is changing. Recent data indicate an increasing number of intra- and inter-sectoral R&D alliances. Anecdotal information suggests that an increasing percentage of cutting-edge research is multidisciplinary in nature. New fields, such as biotechnology and information technology, have changed our perceptions of the way scientific research is conducted and translated into innovation.

In light of these developments, SRS asked the National Research Council (NRC) to undertake a review of its portfolio of data collection and analysis activities and to assist SRS in revising these activities to better meet the information needs of policymakers, managers, educators, and researchers. In response, the NRC created the Committee to Assess the Portfolio of the Science Resources Studies Division of NSF to identify gaps in NSF surveys and provide prioritized recommendations for addressing them.

Methods of Study

Given the breadth of its charge and the short time frame for this study, the committee sought to identify key science and technology resource issues to guide its recommendations in an efficient, yet informative manner. The information for this study, gathered in 1998, included: (1) presentations to the committee by senior SRS managers on their operations and data collection challenges, and later, individual interviews with these managers; (2) focus groups with SRS staff; (3) reports of SRS customer surveys undertaken in 1994 and 1996; (4) interviews with 42 individuals in government, industry, and academia to solicit their views on current issues in science and engineering and the data available to address them; (5) a two-day NRC "Workshop on Data to Describe Resources for the Changing Science and Engineering Enterprise;" and (6) the conclusions and recommendations from a number of recent reports on issues in science and technology policy. The committee also looked at recent data trends to substantiate the importance of issues raised from these sources.

Since updating the portfolio of a federal statistical agency to provide relevant data is an ongoing process, the committee examined how the operations of federal statistical agencies, such as SRS, may be structured to produce continuous renewal of data collection activities. We found that while SRS is considerably smaller than other major federal statistical agencies in budget authority, the division carries out each of the major functions of a federal statistical agency: data collection and acquisition, quality control, preparation of standard data tabulations, data analysis, publications, and data dissemination. Thus, we focused our study on standards for federal statistical agency operations regarding data quality, staff expertise, relevance of concepts measured, data linkages, and timeliness that suggest areas of continued improvement for SRS. However, we highlighted the last three—*appropriateness of concepts and their measurement, ability to link data, and data currency*—as critical dimensions of data relevance that allow statistical agencies to best inform policymaking.

Ensuring Relevance and Establishing Priorities

We have two overarching recommendations that should be considered our highest priorities for SRS and these encourage the division to strengthen its dialogue with its stakeholders and increase its interactions with external researchers.

Other recommendations follow that provide an immediate agenda for the dialogue and interactions.

Appropriateness of Concepts and their Measurement

Recommendation 1. To keep its data relevant and maintain data quality and analytic capacity, SRS should adopt a strategy of continuous review and renewal of the concepts it seeks to measure, and revise its survey instruments, survey operations, and data analysis as needed to keep them current. To achieve this, SRS must strengthen the frequency and intensity of its dialogue and interactions with data users, policymakers, and academic researchers and develop internal processes to convert the feedback it receives from these stakeholders into changes in its surveys and analyses. A key element of this strategy is the creation of advisory committees for SRS surveys that would assist SRS in establishing priorities for future change.

To expand the range of surveys that benefit from advisory committees, we strongly recommend the creation of such a committee for the Survey of Industrial Research and Development. We also recommend that the existing Special Emphasis Panel (i.e., advisory committee) for the Doctorate Data Project (DDP) advise SRS on the Survey of Earned Doctorates (SED), the content of all three SRS personnel surveys, and the design of the Scientists and Engineers Statistical (SESTAT) system. This panel already provides SRS advice on the SED and the Survey of Doctorate Recipients (SDR) and should also provide advice on the National Survey of College Graduates (NSCG) and the National Survey of Recent College Graduates (NSRCG).

A statistical agency should work with data users to define the concepts that it will measure to meet users' information needs. These concepts, and how to measure them, should be continuously reviewed and updated as issues change and as analysis reveals alternate measures that better capture information that is useful to constituents. To operationalize this ongoing process of reviewing and updating data concepts, SRS must increase the frequency and intensity of its dialogue and interactions with data users, policymakers, and academic researchers to capitalize on their insights, expertise, and analytic capabilities. To generate direct interaction of this sort, SRS should establish advisory committees for each of its surveys. These committees will help SRS keep survey content up-to-date and establish priorities for future change. Such committees should be constituted or expanded to cover the Survey of Industrial R&D and all three personnel surveys in the SESTAT system. The SRS Breakout Group of the Directorate for Social, Behavioral, and Economic Sciences (SBE) Advisory Committee and the advisory committees for individual surveys should work together to review and assist in the implementation of our recommendations and set priorities among them and other proposals for change. SRS may also generate dialogue and interaction by holding workshops on emerging issues, improving outreach with constituent groups through booths at conferences, disseminating publications more purposefully, enhancing customer service, and administering a periodic customer survey. SRS should convert the feedback it receives from stakeholders in these ways into means for producing the information its constituents seek. SRS may revise survey instruments, add special modules to instruments to collect data for one survey cycle, establish quick response panels, or employ or sponsor qualitative research as a complement to periodic surveys in order to obtain information more rapidly or more deeply on poorly understood issues.

Data Analysis

Recommendation 2. SRS should more actively engage outside researchers in the analysis of SRS data on current science and engineering resources issues. This may be accomplished by allowing researchers to work at SRS as visiting fellows and by establishing an external grants program. SRS should also monitor and summarize research using its data.

Since the division is limited in its staff size, it cannot have expertise in the full range of subject areas upon which it may be called for data and analysis. SRS and its constituents would benefit from a more interactive relationship between SRS and external researchers, such as university-based researchers, who focus on science and engineering resources issues. The division would expand its analytical range and promote data use by bringing researchers into SRS through a visiting fellows program and by providing grants to researchers who utilize SRS data. SRS should also monitor and summarize research using its data, particularly for its personnel surveys (the National Survey of College Graduates, the National Survey of Recent College Graduates, and the Survey of Doctorate Recipients), which are currently underutilized.

Recommendation 3. SRS and the National Science Board (NSB) should develop a long-term plan for Science and Engineering Indicators *so that it is smaller, more policy focused, and less duplicative of other SRS publications to free SRS resources for other analytical activities.*

NSB and SRS should develop a long-term plan for restructuring *Science and Engineering Indicators*. We believe that *Science and Engineering Indicators* should be smaller and more policy focused. *Indicators* would have more impact on science and technology policy if it focused on bringing analysis to a small set of indicators on issues driving the future of the science and engineering enterprise. There should be a sharper division between the work of a policymaking body such as the National Science Board and the work of a federal statistical agency such as SRS. *Indicators* is redundant with other publications of SRS data which could be referenced in *Indicators* and also linked via hypertext when published on the Internet. Substantial SRS resources—especially staff resources—which are now devoted to the production of this volume, would be freed for other analytic activities if the report were refocused.

engineering. SRS should continue to investigate discrepancies in survey results among its R&D funding and performance surveys and implement changes in survey instruments and operations to address and resolve them. The division should also consider establishing a committee to develop a design for a new, integrated R&D data system that would also account for the apparent increase in the number of intra- and inter-sectoral R&D partnerships. SRS must also develop means for linking its R&D investment data and its human resources data, and help coordinate the data gathering activities of others to improve data availability. Finally, SRS should seek effective ways to allow researchers to link its data with other data sources, public and private.

Data Comparability and Linkages

Recommendation 4. SRS should increase the analytic value of its data by improving comparability and linkages among its data sets, and between its data and data from other sources. Standardizing its science and engineering field taxonomies and other questions across survey instruments is a critical step in this process. Resolving discrepancies in results from different surveys is another.

The ability to link data sets increases the breadth and depth of data, and thereby the ability of analysts to use them to address current issues. SRS's portfolio of data collection activities has been established over the past half century incrementally as individual surveys have been established to provide information on specific pieces of the science and engineering enterprise. To increase the analytic power of its data, SRS should find ways to integrate its data sets further. SRS should continue to improve comparability in questions and response categories across surveys, particularly with regard to questions on field of science and

Data Currency

Recommendation 5. SRS must substantially reduce the period of time between the reference date and data release date for each of its surveys to improve the relevance and usefulness of its data.

In the case of SRS, the timeliness with which data are released is determined by the time that elapses between the reference date in each survey and the date on which survey data are released. The consensus of SRS's data users is that this period of time is generally too long for SRS surveys and that SRS needs to address the length of time taken for survey follow-up, data processing, and the public release of data. To improve the currency of its data, SRS must continue its recent efforts to substantially reduce the period of time between the reference date and data release date for each of its surveys. Means for accomplishing this goal include using incentives for timely response, increased use of the Internet for data collection, and early release of key indicators.

SRS as a Statistical Agency

Recommendation 6. SRS should be seen as a federal statistical agency, and should be supported in its efforts to meet fully those standards set for federal statistical agencies for independence, professional staffing, data quality, and data analysis.

Recommendation 7. SRS's budget is substantially smaller than those of other federal statistical agencies and may need to be increased given the growing importance of its subject area and our recommendations for new processes, data collection activities, and additional studies. Any budgetary increase must be based on clearer information from SRS on its allocation of internal staff and financial resources across its surveys and other activities and on a clearer sense of priorities among current and future surveys and activities, as developed in coordination with advisory committees for its individual surveys and the SRS Working Group of the SBE Advisory Committee.

NSF should see SRS as a federal statistical agency and support the division's efforts as it strives to fully meet standards set for federal statistical agencies regarding independence, professional staffing, data quality, and data analysis. As an important component of this, NSF should provide SRS with additional staff by increasing the number of full-time equivalent (FTE) positions allocated to the division.

SRS's budget is substantially smaller than those of other federal statistical agencies and may need to be increased given the growing importance of its subject area and our recommendations for new processes, data collection activities, and additional studies. We did not have access to sufficiently detailed budget data to conduct a cost-benefit analysis either of the items we recommend or of existing components of the SRS portfolio. Thus, we have not been able

to further prioritize our recommendations, nor have we been able to suggest trade-offs between new and existing activities.

Improving Data Relevance

Science and engineering, a $247 billion[1] a year enterprise of individuals and institutions who carry out research and development in the United States, plays a central role in the advancement of our knowledge-intensive economy and affects the daily lives of Americans in myriad ways. To keep its data relevant for answering today's questions on science and engineering resources, SRS needs to keep its data collection portfolio in synchrony with changes in the science and engineering enterprise. SRS should investigate under-addressed issues in graduate education, the labor market for scientists and engineers, and R&D funding and performance, and use the results of these investigations to revise its surveys and analytic activities.

Graduate School and the Transition to Employment

Recommendation 8. SRS should revise its data collection on issues in graduate education and the job market for new Ph.D.s to better address issues on financial support for graduate students, completion of graduate school, and the transition of new Ph.D.s into employment. SRS must carefully study whether fielding a new longitudinal survey of beginning graduate students, now under consideration, is feasible and cost-effective before committing to such a survey. However, the division should revise the Survey of Earned Doctorates (SED) to

---

[1] The aggregate amount spent by the federal government, industry, nonprofits, and others on research and development in the United States in 1999.

include questions on progress toward degree completion and job market experiences and it should seek to assist professional societies and universities in the collection of standardized data on the job market for new Ph.D.s.

In the face of a difficult job market for recent science Ph.D.s in some fields in the 1990s, policymakers, educators, and analysts have expressed concerns about the efficacy of certain types of support for graduate students and about the outcomes of graduate education. To better understand these issues, they have requested additional data on graduate school completion and attrition, career expectations, educational experiences and skills acquired, student financial support, and the effect of these on graduate school and career outcomes. We recommend that SRS analyze and quickly disseminate the retrospective data it collected through the 1997 SDR on the graduate school and job market experience of Ph.D.s who received their degrees between June 1990 and June 1996. These data should address some of the concerns of policymakers, educators, and analysts in this area. We also recommend that SRS revise the Survey of Earned Doctorates to obtain data on progress through graduate school, perhaps by adding a question asking respondents for the date when all Ph.D. requirements except for the dissertation were completed. We recommend that SRS institute ongoing collection of data about the job market experience of Ph.D.s prior to degree receipt by adding questions on this subject to the SED while also continuing to augment its own data collection by assisting others who are collecting data in this area. We do not, however, recommend that the SED, or a longitudinal survey of graduate students, if fielded, be used to collect data on "skills" obtained by graduate students.

SRS is currently in the development stage for a new longitudinal survey of beginning graduate students, designed to be

responsive to calls for additional data on education and job market experiences of graduate students. SRS should carefully consider the feasibility and costs of developing and administering such a survey. Based on our current understanding, we question the wisdom of such a survey. SRS should weigh whether the issues raised warrant ongoing national data collection from graduate students, examine its ability to collect high-quality national data on attrition and packages of financial support through a longitudinal survey, determine the cost-effectiveness of conducting such a survey, and consider alternative sources of data.

The Labor Market for Scientists and Engineers

Recommendation 9. SRS should revise its data collection on the labor market for scientists and engineers to better capture the career paths of scientists and engineers. SRS should fill gaps in existing data on careers by collecting comparative data on the careers of humanities doctorates, and data on the nonacademic careers of scientists and engineers, on science and engineering field of work, and on the international mobility of scientists and engineers. The division should also work with the Special Emphasis Panel for the Doctorate Data Project to address content and design issues for the SESTAT system to be implemented in the next decade.

To facilitate deeper analysis of personnel issues, SRS should first modify its approach to the design and use of its personnel surveys. SRS should provide better career path data to analysts by making them available at a more detailed level by field. SRS should consider the options available for improving fine field analysis that is currently obstructed by small sample sizes for its personnel surveys. On a related note, science and engineering field of work, dropped from the SDR in 1993, should be restored to the questionnaire. SRS should also exploit the longitudinal nature of its personnel surveys obtained at great expense and with a respondent burden that is difficult to justify if the data are not used longitudinally. Finally, SRS should explore opportunities for linking its personnel data to other career and productivity data, such as data sets on federal research grants, patents, and publications.

SRS should also revise its personnel surveys to better capture the career paths of scientists and engineers. It should consider adding questions to the SDR to obtain additional data on the careers of Ph.D.s who work for government agencies, private businesses, and nonprofit organizations. Such questions might focus on non-salary compensation; patenting and other productivity measures in the private sector; use of scientific background in sales, regulation, or patent law positions; and temporary work arrangements, such as contracting and consulting. SRS must also work with the National Endowment for the Humanities (NEH) and other funding sources to re-institute the humanities component of the SDR, discontinued following the 1995 survey cycle due to budget cuts at NEH. Finally, SRS should augment the data its collects on the global dimensions of the science and engineering workforce by expanding data on foreign scientists and engineers in the United States at all levels—students, postdoctorates, and employees.

R&D Funding and Performance

Recommendation 10. SRS should revise the data it collects on R&D funding, performance, outputs, and outcomes to improve comparability across surveys and to address structural changes in the science and engineering enterprise. SRS should begin by addressing structural changes in industrial research and development, the

relationship between R&D and innovation, the apparent increase in intra- and inter-sectoral partnerships and alliances, and claims that interdisciplinary research is increasing. SRS should examine the costs and benefits of administering the Survey of Industrial Research and Development at the line of business level. SRS should also revise its surveys to address new concepts (e.g., the federal science and technology budget), discrepancies in results among R&D surveys, and the need to obtain better data on academic R&D facility costs.

SRS should improve the accuracy and augment range of its data on industrial R&D and innovation. Currently the Survey of Industrial Research and Development (RD-1), which is administered at the firm level, attempts to disaggregate both applied research and development by asking respondents to distribute these by product group. Firms, however, often ignore this question and the low response rate to product group has made the collected data of little use. We recommend eliminating the product group question from RD-1. As an alternate strategy for obtaining finer detail on industrial R&D, SRS should examine the costs and benefits of administering RD-1 to business units instead of firms. Currently all R&D conducted by a firm is attributed to the firm's predominant industrial category. In an economy dominated by large, multi-product firms, line of business reporting, if feasible, may improve data by obtaining finer detail by industrial classification and geographic location. Also, current R&D expenditure data do not provide adequate information on many activities contributing to innovation. SRS should pursue plans to develop a survey of industrial innovation that addresses the manner in which science and technology are transferred among firms and transformed into new processes and products. SRS should include potential respondents and data users in the development of the survey instrument. SRS should also sponsor research on the nature of

R&D in the service sector; the web of partnerships among firms, universities, and federal agencies and laboratories in conducting R&D; and the extent and role of multidisciplinary research in science and engineering. Results of these investigations should guide SRS in revising its surveys to obtain more complete detail on their role in the science and engineering enterprise. SRS should take steps to better support analysis of the "federal science and technology budget" (FS&T) by requesting that the Department of Energy and the National Aeronautics and Space Administration specify the FS&T portion of their aggregate budget and obligation figures as does the Department of Defense. SRS should continue to investigate and reconcile discrepancies in R&D funding data obtained by its different surveys that hamper analyses of federal R&D funding. It should also complete proposed changes to its Survey of Scientific and Engineering R&D Facilities at Colleges and Universities to provide better data for assessing overhead rates at research universities and estimating future academic infrastructure needs.

1

Introduction

The Division of Science Resources Studies (SRS) of the National Science Foundation is one of fourteen agencies in the federal statistical system, as represented on the Interagency Council on Statistical Policy. Like these other federal statistical agencies—such as the Bureau of Labor Statistics, the Bureau of the Census, the National Center for Education Statistics, and the National Center for Health Statistics—SRS is charged with providing data and analysis on a broad policy area for public- and private-sector constituents. The activities of SRS, authorized by the National Science Foundation Act of 1950 as amended, are important for informing science and technology policymakers, managers, educators, and researchers.

To meet the information needs of its constituents in science and engineering, SRS collects and acquires data on national patterns of research and development (R&D) funding and performance and on the education and careers of scientists and engineers. The division publishes tabulations and analyses of data in these areas and makes its data sets publicly available for analysis by others. SRS also supports the National Science Board in the production of the biennial *Science and Engineering Indicators*. This latter volume, a congressionally-mandated report, is a key reference for science and technology policymakers and analysts.

Federal statistical agencies must continuously monitor developments in the broad subject area about which they collect data to ensure that they are providing information that addresses policy and research needs. Substantial recent developments in both the science and technology policy context and the science and engineering enterprise suggest that the need for SRS to review and update its portfolio is especially important at this time. These trends, some of which have been in progress since the early 1980s, include:

- An increase in industry's share of R&D funding in the United States from about half in 1980 to almost two-thirds today. This shift is largely due to decreases in federal funding for development activities and a high growth rate in industrial R&D spending.

- An increase in service sector R&D which made up just 5 percent of industrial R&D in the early 1980s and now accounts for almost a quarter of it.

- The role of technological innovation in the apparent resurgence in American industrial performance during the 1990s though this is no better understood or quantified than the apparent decline in American industrial performance during the 1970s and 1980s.

- Changing patterns of federal research and development spending following the end of the Cold War characterized by decreases in defense R&D and an acceleration of the long-term growth of federal spending on biomedical research.

- A difficult job market for many new science Ph.D.s in some fields in the early 1990s, characterized by increased numbers of recent Ph.D.s in postdoctoral positions and increased numbers of others working outside of their fields.

- Substantial increases in the number of Ph.D.s awarded in the United States since the mid-1980s, due largely to increases in the number of non-U.S. citizens earning Ph.D.s from U.S. institutions.

- An increase in the percentage of Ph.D.s in science and engineering holding positions outside of educational institutions, with more than half of Ph.D.s so employed for the first time in 1991.

- Trends suggesting an increasing number of intra- and inter-sectoral alliances in research and development.

- Anecdotal information suggesting that an increasing percentage of research, especially "cutting-edge" research, is inter- or multidisciplinary in nature.

- The emergence of new fields, such as biotechnology and information technology, that have changed our perceptions of the way scientific research is conducted and translated into practical innovations.

- The emergence of a global economy and the globalization of the science and engineering enterprise.

In light of these and other developments and their potential for raising important policy issues, SRS asked the National Research Council (NRC) to undertake a review of the SRS portfolio of data collection, acquisition, and analysis activities and to assist SRS in revising these activities to meet the information needs of policymakers, managers, educators, and researchers. The scope of this study was developed through discussions among SRS, the NRC's Office of Scientific and Engineering Personnel, and the NRC's Committee on National Statistics.

Scope of Study

The Committee to Assess the Portfolio of the Science Resources Studies Division of NSF was charged with identifying gaps in NSF surveys and providing prioritized recommendations for addressing those data and information gaps. Potential means for filling these gaps could include recommending new questions for existing surveys, advocating the development of new surveys, suggesting ways for SRS to combine its data with those from other sources in creative ways, and proposing new approaches to data analysis that would benefit those who need these data and information. Other ways to address such gaps include in-depth study of emerging issues to better understand these phenomena and the data about them that would be useful to collect in the future. The committee was also charged with looking for SRS activities that are obsolete and could be discontinued.

The committee convened for this study included experts in a range of fields. The committee members brought expertise in R&D economics, labor economics, graduate education, statistics, and survey methods. They also brought experience in federal science and technology policymaking, managing a federal statistical agency, administration of graduate education, directing industrial R&D, and overseeing university-sponsored research programs. The disciplinary backgrounds of the committee included social sciences, life sciences, physical sciences, and statistics.

Methods of Study

Given the breadth of its charge and the short time frame for this study, the committee sought to identify key science and technology resource issues to guide its recommendations in an efficient, yet informative, manner. The input for this study, gathered in 1998, included:

- presentations to the committee in April 1998, by the SRS Division Director and senior SRS managers on their operations and data collection challenges

- individual interviews with these managers

- focus groups with SRS staff in the Research and Development Statistics Program, the Human Resources Statistics Program, the Science and Engineering Indicators Program, the Integrated Studies Program, and the Information Services Group

- the reports of SRS customer surveys in 1994 and 1996

- interviews with 42 individuals in government, industry, and academia to solicit their views on current issues in science and engineering and the data available to address these issues

- a two-day NRC conference entitled "A Workshop on Data to Describe Resources for the Changing Science and Engineering

Enterprise," held in Washington, D.C. on September 18-19, 1998

- the conclusions and recommendations from a number of recent reports on issues in science and technology policy

In addition, the committee relied on its own members with extensive experience in research and development, graduate education, and the analysis of science and engineering labor markets to assess the importance of various issues raised in the presentations, focus groups, interviews, the two-day conference, and national reports. The committee also looked at recent data trends to substantiate the importance of issues raised. It then selected for treatment in this report those issues raised through this process that it believes are of the highest priority for SRS to address.

The process of updating the portfolio of a federal statistical agency is an ongoing and iterative process. The committee, therefore, also set out to provide SRS with guidance on how to organize both for investigating the issues identified as discussed above and for continuous renewal thereafter. Despite a long history of federal data collection, dating back to the provision for a decennial census in the U.S. Constitution, the literature on criteria for the effective operation of federal statistical agencies is small. The committee found the Committee on National Statistics' *Principles and Practices for a Federal Statistical Agency* useful in outlining issues in this area (NRC 1992). However, the literature specifically on the relationship between a federal statistical agency and its data users in the process of revising and updating a portfolio of data collection and analysis activities is even smaller. Here, the committee relied on the insights of its members who have served in or worked with federal statistical agencies, as well as those of its members who have substantial experience as users of federal statistical data. The committee selected as its highest priorities for SRS two recommendations that focus on strengthening the dialogue and interactions between SRS and its data users.

The Report

In Part I of this report, the committee discusses SRS as a federal statistical agency. Chapter 2 reviews the evolution of the SRS portfolio and relates its current operations to standards for federal statistical agency practices. In Chapter 3, we provide recommendations to SRS for a process that will promote an ongoing interaction between SRS and its data users, allowing SRS to harvest information on the issues for which its customers need data. This process is designed to allow SRS to continuously monitor and revise its portfolio. The committee also recognizes that in some instances, better integration of existing survey data rather than collection of new data may be the most effective way to meet unfilled data needs. Thus, we strongly recommend that SRS seek

to link data sets where possible. We also discuss in Chapter 3 the need for SRS to improve the timelinesss with which it makes its data available to the public.

In Part II, the committee addresses current issues in graduate education, the science and engineering workforce, and research and development funding and how additional understanding of or new data on theses issues could support policymakers. Chapter 4 discusses current policy issues regarding graduate education, the transition to employment for new Ph.D.s, career paths of scientists and engineers, and the international flow of scientists and engineers. Chapter 5 discusses the changing organization of R&D in the United States and globally and then examines specific issues dealing with industrial research and development and the allocation of federal resources for science and technology.

A summary of conclusions and recommendations is presented in Chapter 6.

Part I

SRS as a Statistical Agency

2

History and Assessment of SRS

Evolution of SRS Data Programs

Federal data collection programs date from the first decennial census, required by the U.S. Constitution and conducted in 1790. Federal statistics expanded during the nineteenth century to include agricultural statistics beginning in 1840, income statistics in 1866, education statistics in 1867, and labor statistics beginning in 1884. In the twentieth century, the federal statistical system evolved further. The Bureau of the Census and the Bureau of Labor Statistics (BLS) became bureaus in the new Department of Commerce and Labor in 1903 and BLS was later transferred to the new Department of Labor in 1913. The collection of health statistics was added to the federal statistical portfolio in the early twentieth century and later evolved into the National Center for Health Statistics. Since World War II, policy concerns have led to the creation of the Bureau of Economic Analysis (as the Office of Business Economics in 1953), the Energy Information Agency (1977), the Bureau of Justice Statistics (1979), and the Bureau of Transportation Statistics (1991) (NRC 1997).

The federal collection and dissemination of information on science and technology resources has grown with the federal role in our nation's research and development (R&D) activities. The National Science Foundation Act of 1950 created the Foundation that year with a mission "to promote the progress of science; to advance the national health, prosperity, and welfare; and to secure the national defense." That Act also made NSF a statistical agency. As amended, it requires NSF:

> to provide a central clearinghouse for the collection, interpretation, and analysis of data on scientific and engineering resources, and to provide a source of information for policy formulation by other agencies of the federal government.

These activities were initially carried out by staff in NSF's Office of the Director until the Division of Science Resources Studies (SRS) was created to carry out this mandate. SRS is a division within the NSF Directorate on Social, Economic, and Behavioral Sciences (SBE).

17

Today, SRS is organized functionally into data collection, analysis, and dissemination units. As shown in Figure 2-1, the SRS Director is assisted by a Deputy Director and a Chief Mathematical Statistician. The chief statistician is responsible for setting statistical standards for data collection and analysis across SRS activities and for working to enhance the performance of the division and its staff with respect to these standards. The Human Resources Statistics Program (HRS) and the Research and Development Statistics Program (RDS) have responsibility for primary data collection and acquisition within SRS on personnel, education, and R&D funding and performance. HRS and RDS also produce data tabulations and reports for dissemination to the public. Two additional units focus primarily on analysis activities. The Science and Engineering Indicators Program is responsible for coordinating SRS activities in support of the National Science Boards' biennial publication of *Science and Engineering Indicators,* as well as periodic reports on international comparisons for science and technology resources indicators. The Indicators program also administers a survey on public attitudes toward science and technology. The Integrated Studies Program has responsibility for producing special analyses on science and engineering resource topics of current significance to policymakers and the scientific community. Integrated Studies has, for example, recently published a series of "Issue Briefs" on a variety of personnel and R&D issues. The Integrated Studies Program has also been charged with developing a new survey on innovation. The division is supported by the Information Services Group that oversees publication of SRS reports, both in hard copy and via the SRS web site.

Figure 2-1 Organization of the Division of Science Resources Studies.

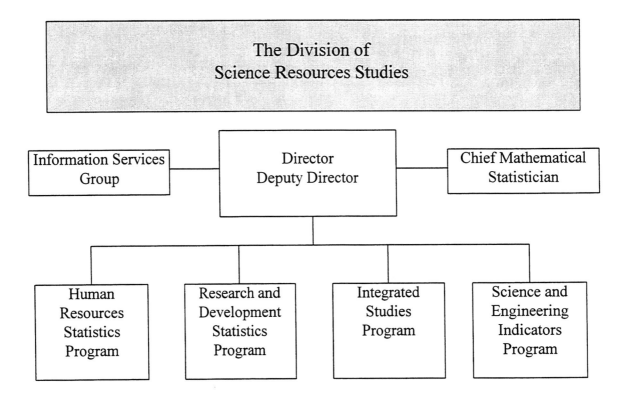

SRS Data Collection Programs

The data collection, acquisition, and analysis activities that SRS administers have grown over the past half century to address the changing data and information needs of policymakers, managers, educators, and researchers in the science and technology policy arena. The science and engineering enterprise was significantly reshaped during and following World War II to include a substantial public role to meet the nation's defense needs, achieve national science and technology goals, and improve the nation's health. To make informed decisions within this new context, federal policymakers required data on and analysis of science and engineering resources issues. To meet these needs, NSF launched its first data collection activities on science and engineering personnel and R&D funding in the early 1950s. As policymakers confronted new issues and problems, the portfolio of data collection activities in SRS grew. Today, SRS data are widely used, yet mostly to address specific issues in narrow topic areas rather than broader questions concerning the level and allocation of resources to the science and engineering enterprise. For SRS to meet these broader needs, it will need to make further efforts to link and integrate its data sets.

Statistics on Science and Engineering Human Resources

Collection of data on Ph.D. scientists and engineers first emerged from a conversation in 1945 between Vannevar Bush, then director of the U.S. Office of Scientific Research and Development, and M.H. Trytten, director of the National Research Council's Office of Scientific Personnel. Drs. Bush and Trytten believed that many of the Ph.D. scientists who had distinguished themselves during World War II had received their undergraduate degrees from small colleges. Trytten proposed to collect data on the undergraduate mentors of Ph.D. scientists to recognize "those devoted individuals who have done the outstanding work of teaching in the sciences in colleges...where the facilities at their disposal are notably meager" (NRC 1951). With initial funding from the Office of Naval Research, the NRC undertook this effort. Trytten's office found that while it was difficult to precisely identify professors who mentored Ph.D. scientists when they were undergraduates, useful data could be obtained on the colleges and universities that awarded bachelor's degrees and Ph.D.s to research doctorate holders. This effort led to the fielding, in 1957, of the NSF-funded Survey of Earned Doctorates (SED)—now a staple of the SRS Human Resources Statistics program. The SED collects data on the field of study, demographic characteristics, educational history, and future plans of new doctorate recipients.

Additional data on graduate students and postdoctoral fellows has been collected since 1966 through the Survey of Graduate Students and Postdoctorates in Science and Engineering (GSPSE). This survey obtains data from colleges and universities on the number of enrolled graduate students in science and engineering by institution and their distribution by field, enrollment status (part-time/full-time), level (first year graduate student, other graduate student, postdoctoral fellow), financial support, and demographic characteristics.

In the act establishing the National Science Foundation in 1950, the U.S. Congress recognized the importance of maintaining current data on the nation's pool of scientists and engineers. In the NSF Act, Congress directed the Foundation to maintain "a register of scientific and technical personnel and in other ways provide a central clearinghouse for information covering all scientific and technical personnel in the United States, including its territories and possessions." In response to this directive, the NSF worked in 1952 with the Federal Security Agency to develop the National Register of Scientific and Engineering Personnel (NRSEP) and then took over its maintenance in 1953. The purpose of the Register, in part, was to monitor the nation's supply of scientific and engineering personnel in preparation for a "national emergency." The NRSEP contained information about the location, fields of specialization, and work activities of scientists and engineers trained to different degree levels (NRC 1989).

In the 1970s, NSF replaced the National Register with a data system based on periodic sample surveys of the nation's scientific and engineering personnel. In response to calls for a more comprehensive look at the nation's science and engineering personnel resources, NSF had, along with other agencies, sponsored a Postcensal Manpower Survey[1] in 1962. A decade later, in 1972, NSF sponsored another postcensal survey, the Professional, Technical, and Scientific Manpower Survey, to provide core data on scientists and engineers in lieu of the discontinued NRSEP. This was supplemented, in 1974, with a New Entrants Survey, designed to obtain data on those who had obtained science and engineering degrees in the United States since the 1970 Census. To cover the population of Ph.D.-level scientists and engineers, NSF also sponsored the first Survey of Doctorate Recipients (SDR) in 1973. In the 1980s, NSF created the Scientific and Technical Personnel Data System (STPDS), drawing on these three surveys.

In the late 1980s, NSF asked the NRC's Committee on National Statistics to convene a study panel to explore the national need for and future characteristics of a data system on science and engineering personnel. In its 1989 report, *Surveying the Nation's Scientists and Engineers*, this panel recommended that the three personnel surveys be better integrated (NRC 1989). Following the recommendations of this report, SRS transformed STPDS into the integrated Scientists and Engineers Statistical Data System (SESTAT), which improved the data quality and integration of the three surveys by modifying sample design, data collection procedures, and questionnaire content for each survey. SESTAT draws data from the National Survey of College Graduates (a postcensal survey), the Survey of Recent College Graduates (a new entrants survey), and the SDR. Question wording on the SDR was revised during the 1993 expansion of the SDR questionnaire to improve comparability of SDR data with data in the rest of the system (Cox, Mitchell, and Moonesinghe 1998b).

While the integration of the content and design of the three personnel surveys in SESTAT has been an impressive success, HRS could undertake further work to improve the comparability and integration of data in its entire set of surveys. For example, the SED and SDR both collect data from doctorate recipients on type of employer, work activities, number of dependents, and disability status but the questions and response categories differ substantively on the two surveys. The addition of a question on starting salary to the SED and a question on current field of science or engineering to the SDR would enhance the complementarity of the two data sets.

Today, as shown in Box 2-1, the Human Resources Statistics Program (HRS) is responsible for data on graduate education for scientists and engineers and data on the nation's science and engineering workforce. HRS conducts the two annual surveys that track science and engineering enrollments and degrees at the graduate level: the Survey of Earned Doctorates (SED) and the Survey of Graduate Students and Postdoctorates in Science and Engineering (GSPSE). HRS also collects workforce statistics on scientists and

---

[1] A postcensal survey uses the U.S. census as a sampling frame to then obtain more detailed data on respondents.

20

engineers derived primarily from the three biennial personnel surveys that the SESTAT system draws on. Further, HRS acquires data on occupational employment from the U.S. Bureau of Labor Statistics and uses data from surveys conducted by the National Center for Education Statistics, including the Integrated Postsecondary Education Data System (IPEDS) Completions Survey, the National Postsecondary Student Aid Survey, and the National Postsecondary Faculty Survey. The Integrated Studies Program also acquires data on the immigration of scientists and engineers from the U.S. Immigration and Naturalization Service.

Box 2-1. Principal Surveys in the SRS Human Resources Statistics Program.

The annual **Survey of Graduate Students and Postdoctorates in Science and Engineering** collects data from all institutions in the United States offering postbaccalaureate programs in science and engineering (11,597 graduate departments in 601 institutions in 1997). Information is collected on student status, demographic characteristics, and major source of federal support. In 1994 the universe comprised doctorate-granting and master's-granting institutions.

The annual **Survey of Earned Doctorates** is a census of all individuals who earn research doctorates from U.S. institutions each school year. It collects information about the field of study, demographic characteristics, educational history, financial support during graduate school, and immediate career plans of these new Ph.D.s. There were 42,683 such research doctorate recipients from July 1997 to June 1998.

The biennial **National Survey of College Graduates**, based on a sample of about 50,000 individuals who reported on their 1990 Census returns that they had at least a bachelor's degree in science and engineering, collects data on the careers of these individuals.

The biennial **National Survey of Recent College Graduates** (also known as the New Entrants Survey) uses a 2-stage probability sample of approximately 25,000 individuals who have received science or engineering bachelor's and master's degrees in the years since the 1990 Census to track the early development of their science and engineering careers.

The biennial **Survey of Doctorate Recipients**, which samples about 50,000 science and engineering doctorate holders, collects longitudinal data on their professional careers.

SRS also obtains data from the Bureau of Labor Statistics' **Occupational Employment Survey.**

Source: http://www.nsf.gov/sbe/srs/educatio.htm.

Research and Development Statistics

The Research and Development Statistics Program (RDS) is responsible for surveys, studies, reports and analyses on the funding and performance of research and development (R&D) in the United States. (See Box 2-2 for definitions of research and development.) RDS conducts surveys on R&D funded and performed by government, industry, universities, and nonprofit organizations. (See Box 2-3.)

Evolution of the Data Collection Activities of the
SRS Human Resources Statistics Program

1945	Vannevar Bush meets with M.H. Trytten of the NRC's Office of Scientific Personnel and asks for data on the B.A. origins of Ph.D. recipients—leads to the NRC's Doctorate Records Project
1950	Congress establishes the National Science Foundation and directs it to maintain the National Register of Scientific and Engineering Personnel (NRSEP)
1952	Federal Security Agency develops the NRSEP
1953	NSF takes responsibility for operating the NRSEP
1957	Doctorate Records Project establishes the annual *Survey of Earned Doctorates* with funding from NSF and other agencies
1962	NSF and other agencies sponsor the Postcensal Manpower Survey
1966	Annual *Survey of Graduate Students and Postdoctorates in Science and Engineering* established
1966	National Center for Education Statistics begins *Completions Survey* as part of its Higher Education General Information Survey
1968	Start of trend data available on *Immigrant Scientists and Engineers* from INS records
1970	Last registration for the National Register of Scientific and Engineering Personnel (NRSEP) held
1972	NSF conducts the *Professional, Technical, and Scientific Manpower Survey* providing core data on scientists and engineers in lieu of the discontinued NRSEP
1973	*Survey of Doctorate Recipients* first fielded to cover the doctoral portion of the NRSEP
1974	*New Entrants Survey* designed to supplement the *Manpower Survey* begun
1975	*Survey of Graduate Students and Postdoctorates in Science and Engineering* begun
1977	Data first available from BLS's triennial *Occupational Employment Statistics Survey*
1982	*Postcensal Survey of Scientists and Engineers* fielded, providing core of the Scientific and Technical Personnel Data System (STPDS)
1982	Congressionally-mandated report *Women and Minorities in Science and Engineering* first published (later expanded to include women, minorities, and persons with disabilities)
1993	The STPDS is replaced by the Scientists and Engineers Statistical Data System (SESTAT)—with data from the National Survey of College Graduates, the Survey of Recent College Graduates, and the Survey of Doctorate Recipients

---

The foundation for the RDS program was laid in the 1950s when NSF first collected data on federal R&D funding and R&D performance. In 1953, NSF established the Survey of Federal Funds for Research and Development, which collects data on R&D obligations made by federal agencies. NSF also began to collect data on R&D performance in 1953 when it funded the first Survey of Industrial Research and Development. The Bureau of Labor Statistics (BLS) fielded the first Industrial R&D Survey for NSF; administration of the survey was later transferred to the U.S. Census Bureau. The same year, NSF conducted the first of six occasional surveys of R&D performance by nonprofit institutions. In 1954, NSF conducted the first of three occasional small-scale surveys of R&D at major universities.

In the 1960s and 1970s, NSF expanded the data it collected on public support for science and engineering. First, NSF deepened the data collected on federal R&D spending. The Survey of Federal Funds for R&D was made stronger by expanding detailed fields of science and engineering in 1960. In 1973, federal obligations for research to universities and colleges by agency and detailed field were also added to this survey. Collection of R&D funding by budget function was also expanded in 1961. Second, NSF established the congressionally-mandated Survey of Federal Support to Universities and Colleges in 1965 to better understand the role of the federal government in supporting academic research and development. Selected information on nonprofit organizations was added to that survey in 1968. Third, to round out data on public support, NSF conducted a survey of R&D support by local governments in 1966 and by state governments in 1967. Other occasional surveys of state support for R&D have also been conducted since.

Box 2-2. Definitions of Research and Development

The National Science Foundation uses the following definitions in its research and development funding and performance surveys:

Basic Research The objective of basic research is to gain more comprehensive knowledge or understanding of the subject under study, without specific applications in mind. In industry, basic research is defined as research that advances scientific knowledge, but does not have specific immediate commercial objectives although it may be in fields of present or potential commercial interest.

Applied Research Applied research is aimed at gaining the knowledge or understanding to meet a specific, recognized need. In industry, applied research includes investigations oriented to discovering new scientific knowledge that has specific commercial objectives with respect to products, processes, or services.

Development Development is the systematic use of the knowledge or understanding gained from research directed toward the production of useful materials, devices, systems, or methods, including the design and development of prototypes and processes.

Source: National Science Board, *Science and Engineering Indictors—1998*, 4-9.

At the same time, NSF expanded and regularized the data it collected on R&D performance. Adding to the Survey of Industrial Research and Development already in the field, NSF began fielding a Survey of Science and Engineering Activities at Universities and Colleges in 1964. This was replaced in 1972 by the annual Survey of Research and Development Expenditures at Universities and College. Also, from 1970 to 1991, SRS had an Industrial Panel on Science and Technology with about 80 members to obtain quick, qualitative information on issues in industrial R&D.

Congress mandated the collection of additional data on R&D performance and infrastructure in the 1980s and early 1990s. NSF first fielded the congressionally-mandated National Survey of Academic Research Instruments and Instrumentation Needs in 1983. In 1986, NSF also implemented the first of what is now the congressionally mandated Survey of Science and Engineering Research Facilities at Colleges and Universities. In 1990, Congress also required NSF to maintain the Master List of Federally Funded Research and Development Centers (FFRDCs).

Box 2-3. Principal Surveys in the SRS Research and Development Statistics Program.

The annual **Survey of Federal Funds for Research and Development** collects information about the characteristics and geographic distribution of all federal R&D funding from the approximately 100 federal agencies and subagencies that obligate funds for R&D.

The annual **Survey of Federal Support to Universities, Colleges, and Nonprofit Institutions** collects information about federal obligations to individual academic and nonprofit institutions by the 15 federal agencies that provide virtually all federal R&D funds to academic institutions.

The annual **Survey of Research and Development in Industry** collects information on R&D expenditures and employment of scientists and engineers from a nationally representative sample of about 23,000 companies (starting with the 1992 survey), including both manufacturing and non-manufacturing companies.

The annual **Survey of Research and Development Expenditures at Universities and Colleges** collects data from a sample of about 460 institutions of higher education that grant science and engineering (S&E) degrees and perform a minimum level of separately budgeted R&D.

The biennial **Survey of Scientific and Engineering R&D Facilities at Colleges and Universities** collects and analyzes data on the availability, condition, need, cost, and funding sources of facilities from a sample of 303 research-performing colleges and universities.

Source: http://www.nsf.gov/sbe/srs/rdstati.htm.

In the 1990s, RDS implemented several initiatives designed to improve or expand the data it collects on government, industrial, and nonprofit R&D funding and performance. In 1992, SRS revised its sample design for the Survey of Industrial Research and Development (RD-1) to improve its measurement of industrial R&D. In particular, the sample was increased and modified to better capture R&D performed by small and non-manufacturing firms. In 1994, the Census Bureau conducted for SRS a pilot innovation survey fielded to 1,000 companies. In 1998, SRS fielded the Survey of Science and Engineering R&D Funding and Performance by Non-Profit Organizations. The last survey of this nature had been conducted in 1973. SRS has also recently funded a survey carried out by the Battelle Memorial Institute of state funding for research and development. In 1998, SRS suspended administration of the Survey of Academic Research Instruments and Instrumentation Needs because of low demand for data from this survey and the need to deploy its resources elsewhere.

Each RDS data collection activity was developed to address a narrow topic rather than to serve as a piece of a cohesive R&D funding and performance data system. Recently, SRS has examined how the results from different R&D funding surveys compare and has discovered discrepancies in funding and performance estimates among its surveys. For example, SRS estimates that federal funding for R&D performed by industry is $31.4 billion based on data from the Survey of Federal Funds for Research and Development, but just $23.9 billion based on data from the Survey of Industrial Research and Development. Further work to improve comparability—even the integration—of these surveys would improve their analytic value.

**Evolution of the Data Collection Activities of the
SRS Research and Development Statistics Program**

1953   The Bureau of Labor Statistics conducts the first *Survey of Industrial Research and Development*

1953   Establishment of *Survey of Federal Funds for Research and Development*

1953   SRS conducts the first of six occasional surveys of R&D performance by nonprofit institutions (the last collected 1973 data)

1953   SRS conducts first of three occasional small-scale surveys of science and engineering R&D at major universities

1957   Conduct of the *Survey of Industrial Research and Development* is transferred to the Census Bureau

1960   Expansion of detailed fields of science in the physical and social sciences reported in the *Survey of Federal Funds for R&D*

1961   Expanded collection of *Federal R&D Funding by Budget Function*

1964   Biennial *Survey of Science & Engineering Activities at Universities and Colleges* begins

1964   Congressionally-mandated, annual *Survey of Federal Support to Universities and Colleges* is first fielded

1964   SRS conducts survey of R&D support by local governments

Evolution of the Data Collection Activities of the
SRS Research and Development Statistics Program
(continued)

1964   First of several occasional surveys of R&D support by State Governments (latest survey in 1996)

1967   First known publication of *National Patterns of R&D Resources*

1968   Selected information on nonprofit organizations added to what is now the *Survey of Federal Science and Engineering Support to Universities, Colleges, and Nonprofit Institutions*

1970   *Industrial Panel on Science & Technology* established (approximately 80 members) to obtain qualitative data on Industrial R&D.

1972   Annual *Survey of Research and Development Expenditures at Universities and Colleges* replaces *Survey of Science & Engineering Activities at Universities and Colleges.*

1973   Federal obligations for research to universities and colleges by agency and detailed S&E field added to the *Survey of Federal Funds for Research and Development*

1981   SRS sponsored surveys of industrial innovation begun by John Hansen, Christopher Hill, and James Stein

1983   Congressionally-mandated, triennial *National Survey of Academic Research Instruments and Instrumentation Needs* begun

1983   SRS publishes yearly *Highlights* providing estimates of R&D expenditures in the year ahead; estimates based on input from the Industrial Panel on Science and Technology

1986   Congressionally-mandated *Science & Engineering Research Facilities at Doctorate-Granting Institutions* established (later becomes *Science and Engineering Research Facilities at Colleges and Universities*)

1990   Congress requires NSF to maintain *Master List of FFRDCs (Federally Funded Research and Development Centers)*

1991   Industrial Panel on Science & Technology discontinued

1992   *National Survey of Academic Research Instruments and Instrumentation Needs* becomes biennial

1992   Sample design for the *Survey of Industrial Research and Development* revised to better capture R&D performed by small and non-manufacturing firms

1994   Bureau of the Census conducts a pilot innovation survey of 1,000 companies for SRS

1995   *Survey of Federal Funds* and *Survey of Federal Support* first provide data on the Department of Defense's Science and Technology budget

1997   Formal collection of the *Survey of Academic Research Instrumentation and Instrumentation Needs* suspended

1998   SRS fields the *Survey of Science and Engineering R&D Funding and Performance by Nonprofit Organizations* (last conducted in 1973)

Data Publication, Integration, and Analysis

Data Publication

SRS published about 30 reports in 1998 and about 50 in 1999. These include detailed statistical tables, data and issue briefs, periodic reports, and special reports:

- Detailed Statistical Tables: reports containing an extensive collection of tabulated data from each of SRS's surveys

- *Data Briefs* and *Issue Briefs*: short reports highlighting results from recent surveys and analyses

- Periodic reports, such as *Science and Engineering Indicators; National Patterns of R&D Resources; Women, Minorities, and Persons With Disabilities in Science and Engineering;* and *International Science and Technology Data Update*

- Special reports, such as *Undergraduate Origins of Recent Science and Engineering Doctorate Recipients* and *International Resources for Science and Technology*

In addition to publications, SRS disseminates some of its data in electronic format. SRS has developed a web site, highly regarded by individuals interviewed for this study, that provides on-line access to tabulations, reports, and two SRS databases, WebCASPAR (Computer-Assisted Science Policy Analysis and Research system) and SESTAT (Scientists and Engineers Statistical Data System). WebCASPAR is a database system containing multiyear data on science and engineering resources at individual academic institutions and by field. SRS draws on data for WebCASPAR from several of SRS's academic surveys plus information from a variety of other sources, including the National Center for Education Statistics. SESTAT is an integrated data system that draws on SRS personnel surveys to provide data on the employment, educational and demographic characteristics of scientists and engineers in the United States. WebCASPAR and SESTAT provide examples of ways in which SRS data or data sets can be integrated or linked and future SRS efforts to link its data within and across its programs would facilitate deeper analysis of the rich data sets that SRS has created.

SRS also makes its survey data sets available for researchers to use in studying science and engineering resources issues in further depth, though within the restrictions of confidentiality requirements. Like other federal agencies, SRS is required to comply with the provisions of the Privacy Act of 1974 that require that confidentiality of records on individuals be maintained. Further, SRS also provides its own assurances to survey respondents that data will not be disclosed that would permit their identification. Thus, while SRS may release its data sets to outside researchers, the division strips identifying information (name, address, social security number) from its data sets before doing so and, in some instances, SRS suppresses fields or recodes data to further protect confidentiality. When recoding or field suppression would be so extensive that it is not cost effective to produce a data file, SRS attempts to devise alternate means for providing data to researchers.

Data Analysis

Data analysis in SRS is carried out across the HRS, RDS, Integrated Studies, and Indicators Programs. HRS and RDS program staff prepare data briefs that announce key highlights from surveys at the time survey data are released or shortly thereafter. Staff in these programs also prepare substantial periodic reports, such as *National Patterns of R&D Resources* and *Women, Minorities, and Persons with Disabilities in Science and Engineering* that draw on data from surveys in R&D funding and performance and human resources, respectively. The reorganization of SRS in 1997 created an Integrated Studies Program to promote more integrated analysis of SRS data sets. Staff in this program conduct analyses for the biennial *Science and Engineering Indicators* and for publication in topical *Issue Briefs*. Box 2-4 displays *Issue Briefs* published by SRS through May of fiscal year 1999 by research area. These provide data and analysis on a variety of important topics.

Box 2-4. SRS Issue Briefs Published in Fiscal Year 1999 (through May 1999).

R&D Funding and Performance
- What Is the State Government Role in the R&D Enterprise? (May 26, 1999)
- What is the Federal Role in Supporting Academic Research and Graduate Research Assistants? (April 16, 1999)
- How Has the Field Mix of Federal Research Funding Changed Over the Past Three Decades? (February 17, 1999)
- How Has the Field Mix of Academic R&D Changed? (December 2, 1998)
- What are the Sources of Funding for Academically Performed R&D? (December 23, 1998)
- Venture Capital Investment Trends in the United States and Europe (October 16, 1998)
- U.S. Inventors Patent Technologies Around the World (February 24, 1999)

Science and Engineering Labor Market
- Will Small Business Become the Nation's Leading Employer of Graduates with Bachelor's Degrees in Science and Engineering? (March 4, 1999)
- Degrees and Occupations in Engineering: How Much Do They Diverge? (December 31,1998)
- How Much Does the U.S. Rely on Immigrant Engineers? (February 11, 1999)
- What Follows the Postdoctorate Experience? Employment Patterns of 1993 Postdocs in 1995 (November 27, 1998)

Science and Engineering Education
- Does the Educational Debt Burden of Science and Engineering Doctorates Differ by Race/Ethnicity and Sex? (April 16, 1999)
- Retention of the Best Science and Engineering Graduates in Science and Engineering (February 23, 1999)
- Have Forms of Primary Financial Support for S&E Graduate Students Changed During the Past Two Decades? (December 4, 1998)
- Has the Use of Postdocs Changed? (December 2, 1998)

SOURCE: http://www.nsf.gov/sbe/srs/issuebrf/ib.htm

Science and Engineering Indicators

In addition to compiling data from its surveys and other quantitative data sources for its own publication, SRS assists in the production of the National Science Board's (NSB's) biennial *Science and Engineering Indicators* report. Beginning with *Science Indicators—1972*, NSB has sought to publish a volume that gauges the status of the science and engineering enterprise and its contributions to national goals and the national welfare. Congress later amended the National Science Foundation Act of 1950 to require the NSB "to render to the president, every even-numbered year, a report, for submission to Congress, on indicators of the state of science and engineering in the United States" (NSB 1996). To date, the NSB has published thirteen volumes in this series, the latest being *Science and Engineering Indicators—1998*.

Indicators describes elementary and secondary science and mathematics education, postsecondary science and engineering education, the science and engineering workforce, research and development funding patterns, academic R&D performance, and industrial R&D. The report places each of these subjects in international context. SRS fields, on a biennial basis, a Survey of Public Attitudes and Public Understanding that monitors both public attitudes toward science and technology and the public's level of scientific understanding and policy preferences on selected issues. The results of this survey are published in the *Indicators* report. Each volume also includes a chapter on a special topic. In 1998, *Indicators* included a chapter on the economic and social significance of information technology. The contents of the 1998 volume are outlined in Box 2-5.

Assessing SRS

Data collection activities at NSF have grown over the last half century into the Division of Science Resources Studies—a small federal statistical agency with a range of programs centered on the topic of science and engineering resources. In the 1990s, SRS managers have worked to upgrade statistical standards, build staff expertise, revise surveys to increase data relevance, better coordinate data sets, improve the timeliness of data release, and expand data analysis. While this has been progress, the division's capabilities must be developed further in order to produce substantial benefits for the science and technology policymakers, planners, educators, and researchers.

Operations and Resources

SRS has not historically been viewed as a federal statistical agency by NSF. Yet the SRS staff of about forty are called on to collectively carry out each of the major functions of a federal statistical agency: data collection and acquisition, quality assurance, preparation of tabulations and public use data files, data analysis, publication of reports, and data and report dissemination. Generally speaking, SRS carries out these activities well, but there is room for improvement in statistical quality, staff expertise, data relevance and timeliness, and depth and breadth of analysis. These areas are discussed in the sections below.

While SRS's mandate is narrowly focused around a particular topic, science and engineering resources, its budget of $14.5 million per year (FY 1999 estimated) is small relative to other agencies with similarly

Box 2-5. National Science Board, *Science and Engineering Indicators—1998,* Contents.

Chapter 1: Elementary and Secondary Education
       Student Achievement
       Curriculum and Instruction
       Teachers and the Profession of Teaching

Chapter 2: Higher Education in Science and Engineering
       Worldwide Increase in S&E Educational Capabilities
       Characteristics of U.S. Higher Education Institutions
       Undergraduate S&E Students and Degrees in the United States
       Graduate S&E Students and Degrees in the United States
       International Comparisons of S&E Training in Higher Education

Chapter 3: Science and Engineering Workforce
       Labor Market Conditions for Recent S&E Degree-Holders
       Selected Characteristics of the S&E Workforce
       S&E Job Patterns in the Service Sector
       Scientists and Engineers in an International Context: Migration and R&D Employment
       Projected Demand for S&E Workers

Chapter 4: U.S. and International Research and Development: Funds and Alliances
       National Trends in R&D Expenditures
       R&D Patterns by Sector
       Inter-Sector and Intra-Sector Partnerships and Alliances
       Government R&D Support
       International Comparisons of National R&D Trends

Chapter 5: Academic Research and Development: Financial and Personnel Resources, Integration With Graduate Education, and Outputs
       Financial Resources for Academic R&D
       Academic Doctoral Scientists and Engineers
       Integration of Research with Graduate Education
       Outputs of Scientific and Engineering Research

Chapter 6: Industry, Technology, and Competitiveness in the Marketplace
       U.S. Technology in the Marketplace
       International Trends in Industrial R&D
       Patented Inventions
       Venture Capital and High-Technology Enterprise
       New High-Tech Exporters
       Summary: Assessment of U.S. Technological Competitiveness

Chapter 7: Science and Technology: Public Attitudes and Public Understanding
       Interest in Science and Technology
       Understanding of Scientific and Technical Concepts
       Attitudes Toward Science and Technology Policy Issues
       Sources of Scientific and Technical Information

Chapter 8: Economic and Social Significance of Information Technologies
       Information Technologies
       Impacts of IT on the Economy
       IT, Education, and Knowledge Creation
       IT and the Citizen

specific objectives, such as the Bureau of Justice Statistics. As seen in Table 2-1, the FY 1999 budgets of the other major statistical agencies range in size from $25.0 million in the Bureau of Justice Statistics to $1.35 billion for the U.S. Census Bureau.

In general, NSF should see SRS as a federal statistical agency and should support the division in its efforts to meet fully those standards for statistical agencies regarding independence, professional staffing, data quality, and data analysis. We recommend that NSF provide SRS with additional staff by increasing the number of full-time equivalent (FTE) positions allotted to the division and that SRS actively engage external researchers through visiting fellowships and external grants. Additional staff and the engagement of outside researchers will increase the breadth of activities the division may undertake as well as the depth of skills brought to these activities. Further, since SRS's budget is substantially smaller than those of other agencies with a specific policy focus its resources may need to be increased given the growing importance of its subject area and our recommendations for new processes and data collection activities as discussed in Chapters 3, 4 and 5. Any budget increase, however, must be based on an informed analysis of the allocation of financial and staff resources across SRS activities and on a clear sense of priorities among current and proposed activities.

Table 2-1 Fiscal 1999 (estimated) and 2000 (requested) Budgets for Major Federal Statistical Agencies (millions of dollars)

Statistical Agency	Department or Agency	1999 (estimated)	2000 (request)
Bureau of the Census: current programs	Commerce	156.1	166.9
Bureau of the Census: periodic programs (censuses)	Commerce	1193.8	2914.8*
Bureau of Economic Analysis	Commerce	43.1	49.4
Bureau of Labor Statistics	Labor	398.9	420.9
National Agricultural Statistics Service	Agriculture	104.0	100.6
Economic Research Service	Agriculture	65.8	55.6
National Center for Education Statistics	Education	104.0**	117.5**
National Center for Health Statistics	HHS	94.6	109.6
Energy Information Administration	Energy	70.5	72.6
Bureau of Justice Statistics	Justice	25.0**	32.5**
Bureau of Transportation Statistics	Transportation	31.0	31.0
Statistics of Income Division, Internal Revenue Service	Treasury	28.8	30.9
Science Resources Studies Division	NSF	14.5**	14.9**

* Developed before the Supreme Court decision on sampling.

** Funding levels shown for the National Center for Education Statistics, the Bureau of Justice Statistics, and the Science Resources Studies Division do not include salaries and expenses from other departmental sources.

Source: Council of Professional Associations on Federal Statistics (Spar 1999); National Science Foundation FY 2000 Budget Request.

Statistical Agency Practices

In *Principles and Practices for a Federal Statistical Agency*, the Committee on National Statistics has delineated a useful set of standards against which SRS—as a statistical agency—may be assessed. Here we highlight SRS operations relative to standards for data quality, staff expertise, data relevance, data linkages, and analytical activity to suggest areas that need additional improvement.

Data Quality

The Committee on National Statistics specifies that to assure commitment to data quality a federal statistical agency should:

- develop an understanding of the validity and accuracy of its data and convey the resulting measures of uncertainty and sources of error to users

- undertake ongoing quality assurance programs to improve data validity and reliability and to improve the processes of gathering, compiling, editing, and analyzing data

- use modern statistical theory and sound statistical practice in all technical work (NRC 1992)

Following from these, a federal statistical agency must fully describe its data and comment on their relevance to specific major uses. As described in an NRC report on the Bureau of Transportation Statistics, the agency should "describe the methods used, the assumptions made, the limitations of the data, the manner by which data linkages are made, and the results of research on the methods and data. Measures of uncertainty should be provided to users, and statistical standards should be published to guide professional staff in the agency as well as external users" (NRC 1997a).

SRS has a good track record of improving data quality and meeting statistical standards in the recent past, but should take additional steps to ensure that standards are met across SRS operations. For example, SRS has undertaken substantial work in the 1990s to upgrade data quality for the three personnel surveys in the SESTAT system. In addition, SRS has required contractors to provide detailed methodology reports. To this end, the National Research Council, for example, began producing a methods report for the Survey of Earned Doctorates beginning with the 1990-1991 survey year.

SRS could improve data quality and information about it further by taking at least three additional steps and possibly others. First, it should require that all of its contractors provide methods reports that address data quality standards. The contractor for the Survey of Public Attitudes, for example, does not currently provide a methodology report that details measures of data quality. Second, SRS should continue recent efforts to provide staff with professional development opportunities for improving their statistical skills, so they can better implement quality assurance programs. Third, SRS should continue to develop and strengthen a program of methodological research, undertaking rigorous analysis of the data collected to assess the quality of the data relative to concepts it is supposed to measure.

Professional Staff and External Expertise

Given the small size of SRS and its need to accomplish each of the major tasks of a federal statistical agency, SRS staff are required to perform multiple functions. In larger statistical agencies, each staff person would have a more focused role. In SRS, staff are called on to perform several roles at a time. For example, survey officers not only supervise survey contractors, they also conduct data analyses, write data briefs, and contribute to reports. Likewise, program directors must perform their managerial duties, but also continue to analyze data and write reports.

To meet the expectations of the NSB and other constituents, SRS staff must have adequate expertise in both the theory and practice of statistics and in the disciplines relevant to the analysis of data on science and engineering resources (NRC 1992). As of spring 1999, SRS has a staff of 42, very small compared to other federal statistical agencies whose staff sizes range into the thousands. Many staff have expertise or experience in statistical methods, survey management, economics, science policy, education, and international comparisons of science and engineering. However, staff skills in areas such as statistical methods, generally speaking, should be improved and the range of staff expertise should be expanded.

First, SRS may augment its staff expertise through professional development activities. SRS has begun identifying areas in which staff require additional training to improve statistical methods. These areas could be addressed in-house through individual training, group discussions with staff on statistical techniques (perhaps "brown bag" discussions scheduled during the lunch hour), and review of staff work. Staff may also upgrade skills by attending relevant graduate courses in statistical methods and data analysis, participating in short courses offered through continuing education programs, and staying active in relevant professional associations. Some statistical agencies make explicit their commitment to staff development by specifying professional development goals for each of their staff as part of their performance reviews. Such a process could ensure that staff take necessary steps to acquire and maintain appropriate skill sets (NRC 1997a).

SRS should also develop a staffing plan that allows it to augment staff expertise,

particularly in key areas. Earlier in the 1990s, NSF reduced the number of SRS staff and the staffing level of the division has remained flat since, even though demands on the division have increased. This reduction in SRS personnel compounded other cuts in the resources available to NSF for data analysis. In the 1970s and 1980s much use was made of SRS data in analyses by NSF's Division of Policy Research and Analysis (PRA) and its predecessors. When NSF terminated this office, the resources that had been available to PRA were never fully made available to SRS. Together, these reductions have left NSF and the science and engineering community where they are today—short on staff resources for both data collection and analysis. Within the cap on its number of full-time equivalent (FTE) staff, SRS should use staff turnover as a means to attract skilled staff and meet needs for expertise in statistical methods and important analytical subject areas. New hires in HRS and RDS should have strong skills in statistical methods and subject matter knowledge. New hires in the Indicators or Integrated Studies programs should have strong analytical skills, subject matter knowledge, and the ability to draw on diverse data to provide meaningful analyses of science and engineering resources. We would also like to see NSF allow the number of SRS staff to grow so that the division may better meet its constituents' expectations. This would allow SRS to more rapidly acquire the skilled staff it needs, broaden the range of its activities, and improve the quality of its work.

Finally, because its staff is small, SRS cannot have expertise in all subject areas for which it could conceivably be called upon to provide data and analysis. SRS needs to develop a more interactive relationship with external researchers to increase the division's range of professional capabilities. SRS currently invites outside experts to attend workshops and serve on advisory panels. By also establishing programs to more actively engage these experts, SRS could expand the breadth of research it undertakes and create opportunities for interaction between SRS and external data users, producing insights that would benefit SRS staff, researchers, and the relevance and quality of the data. Several statistical agencies have visiting fellows programs in which distinguished statisticians or other researchers work in the agency for a specified period of time. NSF has a long history of experts rotating into the agency on a temporary basis and this practice could be extended to SRS. Also, many statistical agencies have programs that provide grants to external researchers to use their data. SRS has the authority to administer such a program, but has not issued a general announcement since the mid-1980s because of budget constraints. SRS has awarded external grants since, but only on an ad hoc basis. Grant sizes have ranged generally from $50,000 to $150,000. Today, the division needs to focus additional effort on data analysis and revitalizing a program of external grants would increase the use and analysis of SRS data, especially if targeted toward underutilized data such as those in SESTAT.

Data Relevance

SRS is a federal statistical agency that exists to serve the information needs of policymakers, program administrators, planners, educators, and researchers in the science and engineering community. It should, therefore, meet the standard for data relevance set by the Committee on National Statistics: "a federal statistical agency must be in a position to provide information relevant to issues of public policy" (NRC 1992).

Indeed, SRS's data are useful to its constituents and widely used. SRS data are published in the National Science Board's biennial *Science and Engineering Indicators*, an important reference for federal science and technology policy makers. They are also used widely in policy reports on basic research, graduate education, and scientific and technological workforce issues.

Yet, in the 1990s, changes in the science and engineering enterprise have left SRS lagging behind the data needs of its constituents and issues have emerged for which SRS has not been able to provide adequate data. (Examples of these issues in graduate education, the labor market for scientists and engineers, and R&D funding are detailed in Chapters 4 and 5.) The Committee on National Statistics states that "an agency's mission should include responsibility for assessing the needs for information of its constituents" (NRC 1992). SRS has taken some steps to assess user needs, such as fielding a customer survey on a periodic basis and contracting with the NRC to produce this report. SRS has administered customer surveys in 1996 and 1999 that seek to ascertain customer satisfaction with various aspects of SRS data and publications. However, SRS needs to take additional steps to develop and implement means for reviewing, updating, and supplementing its data collection and acquisition activities on an ongoing basis to meet current information needs.

To keep its data relevant, SRS must substantially and continuously improve and supplement its survey instruments and data analysis. To achieve this, SRS must strengthen its dialogue and interactions with policymakers, academic researchers and other data users to capitalize on their insights, expertise, and analytic capabilities. Means for accomplishing this include establishing advisory committees for each survey; holding a series of workshops on emerging issues in the science and engineering enterprise; improving outreach with constituent groups; more purposeful dissemination of publications; and the promotion of data use through easier access to data and programs to more deeply involve external researchers in SRS data analysis. SRS must also develop internal processes to convert the feedback it receives from stakeholders in these activities into changes in its surveys and issues for analysis. Improvements in data collection almost inevitably come at a higher cost; however, since SRS resources are limited, it is critical that priorities for change be based on both the relevance of the data and the costs involved.

To improve the currency of its data, SRS must also continue its recent efforts to substantially reduce the period of time between the reference date and data release date for each of its surveys. Means for reducing survey processing time and the time spent in releasing data include incentives for timely response, increased use of the Internet for data collection, and early release of key indicators.

Finally, SRS should employ quick response surveys to obtain information more rapidly on pressing issues and use qualitative methods as a complement to periodic surveys in order to more deeply investigate poorly understood issues.

Data Linkages

The Committee on National Statistics states that an effective statistical agency promotes data linkages in order to enhance the value of data sets and their analytic power (NRC 1992). SRS's portfolio of data collection activities has grown over the past half century as individual surveys have been established to provide information on specific pieces of the science and engineering enterprise. SRS has only recently begun to manage its surveys as components of a more integrated data system and would increase the depth and usefulness of its data by pursuing this further. SRS should find ways beyond SESTAT and WebCASPAR to link data sets. For example, creating linkages between its R&D funding and human resources data programs would give a better overall picture of resources available for science and engineering. Standardizing its science and engineering field taxonomies and revising

questions to improve comparability across survey instruments are critical steps that would facilitate this process. SRS should not limit itself to these steps, however, but should also find ways to allow researchers to link its data to those from other data sources, public and private. It should also encourage standardization in university data collection on the career paths of science and engineering graduates and continue to play a lead role in collecting, coordinating, and standardizing international S&E resource data.

Data Analysis

The Committee on National Statistics also stipulates that "an effective statistical agency should have a research program that is integral to its activities" (NRC 1992). Analysis of substantive issues for which data were compiled has two goals. First, it provides information on these issues through issue briefs and reports to policymakers, and other constituents. Second, analysis will indicate limitations in an agency's data and thus guide how its surveys could be redesigned to improve concepts and fill data gaps.

SRS currently provides data analysis through data briefs, issue briefs, periodic reports, and special reports. The public release of SRS survey data is announced through the publication of a data brief that highlights one important trend from the data set and thus seeks to couple the release to an important current issue. SRS has recently increased the number of issues briefs it publishes, seeking to bring SRS data to the analysis of a specific, narrowly focused science and engineering resources issue. SRS also has an important relationship with the National Science Board in the production of a major statistical report, *Science and Engineering Indicators*. SRS is responsible for compiling data from its surveys and other quantitative data sources and producing the *Indicators* report under the guidance and on behalf of the NSB.

SRS should carefully consider how it may best engage in and support research on science and engineering resources in the future, seeking greater analytic use of its data. We support the recent effort of SRS to expand its analytical program and the publication of topic-specific issue briefs. As noted above, we also urge SRS to engage outside experts and researchers in research activities using SRS data. The division has not realized the potential benefits of more fully interacting with outside researchers and leveraging their expertise. With a relatively small investment in grants to outside researchers or a program to bring researchers into SRS on a temporary basis, SRS would benefit from an increase in the breadth and depth of data analysis. Such a program also allows SRS analytical flexibility, as specific researchers may be engaged based on their expertise as substantive issues change.

Finally, NSB and SRS should develop a long-term plan for restructuring *Science and Engineering Indicators*. Individuals interviewed for this study as well as science and technology policymakers recently interviewed by SRS following the publication of the last *Indicators* volume suggest at least three possible futures for *Indicators*. The first of these, of course, is maintaining the status quo. As currently conceived, *Indicators* provides a wealth of information on science and engineering resources in the United States, and increasingly, in an international framework, which benefits both the NSB and SRS. It provides the NSB a means for highlighting important science and engineering issues. It also provides SRS a

means for showcasing its data in a high profile report considered by many an essential reference for quantitative information by science and technology policymakers. The second is for the NSB to reduce the amount of policy analysis in the volume and concentrate on the data presented. Those with this perspective believe that the policy analysis presently in *Indicators* is not very useful, while the data are. The third is for the NSB to make the document more focused on policy issues and less on data. Individuals holding this point of view suggest that *Indicators* would have a greater impact if it were smaller, less redundant with other SRS publications, and offered policy insights built on important indicators.

We believe that *Science and Engineering Indicators* should be smaller and more policy focused. *Indicators* would have more impact on science and technology policy if it focused on bringing analysis to a small set of indicators on issues driving the future of the science and engineering enterprise. There should be a sharper division between the work of a policymaking body such as the National Science Board and the work of a federal statistical agency such as SRS. Much tabular material in *Indicators* is redundant with other publications of SRS data, which could be referenced in *Indicators* and also linked via hypertext when published on the Internet. Substantial SRS resources—especially staff resources—which are now devoted to the production of this volume, would be freed for other analytic activities if the report were refocused.

3

Dimensions of Relevance

The previous chapter outlined areas in which SRS may improve its operations and its processes for ensuring the continuing relevance of its data and analyses. This chapter expands the discussion on ensuring relevance by focusing on three key dimensions—appropriateness of concepts and their measurement, ability to link data, and data currency. Specific substantive areas in which the relevance of SRS data could be improved are discussed in chapters 4 and 5.

Relevance of data has many dimensions and each of these should be considered in determining whether a statistical agency is meeting the needs of its constituents (NRC 1997a). For purposes of this study, we focus on three aspects of data collection that influence the relevance and analytic value of SRS data:

Appropriateness of concepts and their measurement. A statistical agency should, working with data users, define the concepts that it will measure in order to meet their information needs. These concepts and how to measure them should be continuously reviewed and revised as issues change, and as analysis reveals weaknesses in current measures and suggests alternate measures that

might better capture information of use to constituents. Issues of measurement include the ability to quantify these concepts in a meaningful way, the ability to reliably collect data, and the level of detail that both meets user needs and is cost-effective.

Ability to link data. The ability to link data collected through various instruments within a data collection program and to link that program's data to external data sources increases the breadth and depth of data, and thereby, the ability of analysts to use them to address current issues.

Data currency. Whether data reflect current conditions depends on several factors. The first of these is the periodicity of each survey. Data that are collected biennially, for example, are expected to have a shelf life twice that of annually collected data. The second of these is the timeliness with which collected data are released for public use. Timeliness is measured, in the case of SRS, as the time that elapses between the reference date in each survey and the date at which survey data are released. The third factor is the rate at which trends in the field surveyed actually change, and sometimes more importantly, are perceived to change.

A statistical agency may best influence current and future policy debates by improving its operations across all three of these dimensions of relevance. This chapter discusses each of these in turn.

Relevance of Data Concepts

First of all, the concepts measured by a statistical agency must be appropriate and up-to-date. Statistical agencies are charged with providing data and analyses on subject areas relevant to policymakers, researchers, and others with a stake in the targeted policy issues. Agencies must identify the appropriate set of concepts that capture important trends and issues in that subject area and then establish appropriate measures for the concepts, which they apply through their data collection and acquisition activities. These measures should be established at a level of detail that allows data analysis to provide answers to the questions data users are concerned about.

Statistical agencies must establish means for monitoring and updating the set of concepts they measure and their data collection and acquisition activities as their subject areas change. An important source of information for statistical agencies on the continuing relevance of concepts in their specific subject areas is their community of data users. As the Committee on National Statistics has advised, a statistical agency's mission "should include responsibility for assessing the needs for information of its constituents" in order to ensure that the data and information it provides continue to be relevant over time (NRC 1992).

SRS Data Users

While SRS data are widely used, the evolution of the SRS portfolio described in the previous chapter has not been sufficient recently to allow the division to provide all the data needed by national policymakers as they

deliberate questions about resources in a fast-changing world. The science and engineering enterprise has changed substantially in the last two decades. With the exception of the three personnel surveys and some aspects of the Survey of Industrial Research and Development, however, SRS data collection activities have changed little during this time. As a result, policymakers, planners, and researchers have confronted new resource issues for which data do not currently exist.

In planning its program and identifying emerging issues of import, a statistical agency "should work closely with policy analysts in its department, other appropriate agencies in the executive branch, relevant committees and staff of the Congress, and appropriate non-governmental groups" (NRC 1992). For SRS, these constituents include policymakers in the Executive Office of the President, Congressional committees and staff, the National Science Board, and officials in federal agencies. They include policymakers at the state and local level who play a role in education and technology-based economic development. They include those who seek to influence or inform policy, such as the National Academy of Sciences, professional associations, and think tanks. SRS data users also include academic administrators in the nation's colleges and universities and planners and policymakers in industry who have much at stake in federal policies and programs. Academic researchers seeking to understand scientific processes and explore relevant science and technology policy issues are also important data users. The media are also a constituency as they serve as one means for disseminating data and issue-oriented analysis to the SRS audience. Finally, students and faculty are SRS customers and the more informed their career and mentoring decisions, the more effective the science and engineering enterprise.

Use of SRS Data

SRS has served these constituents reasonably well. Many among the people interviewed for this study believed SRS is already doing an excellent job. One individual remarked that overall, SRS data are the "gold standard" for data related to science and technology, and noted that coverage of issues is very good. Another interviewee said that SRS data are very helpful for U.S. science and technology issues: "Their domestic data are essential; they are the official statistics. Their international publications are also useful. The data in such publications as *Science and Engineering Indicators*, *National Patterns of R&D Resources*, and *Federal Funds* are quite detailed and very useful."

While use of data is not an indicator of satisfaction with those data, the widespread use of SRS data is notable. The National Science Board, for example, relies heavily on SRS data as can be seen in *Science and Engineering Indicators*. SRS data are widely used by officials in the U.S. Office of Management and Budget (OMB), the White House Office of Science and Technology Policy (OSTP), and the Congressional Research Service (CRS) to analyze research budgets and science and technology policy issues. These officials are not necessarily intense SRS data users, but when they do use SRS data it tends to be for important federal science and technology policy issues. These officials rely mainly on published materials, particularly *Science and Engineering Indicators* and SRS publications on research and development funding. University administrators are also perennial users of SRS data. Graduate deans widely use data from the Survey of Earned Doctorates and other administrators use SRS data on university R&D and academic facilities. In SRS surveys for which university administrators are respondents, SRS obtains very high response rates—around 95 percent—because the data are, in turn, used by these administrators.

Besides this, SRS data show up in reports prepared by various other organizations. The American Association for the Advancement of Science (AAAS) primarily uses data from the U.S. Office of Management and Budget and from federal agencies in its annual report on federal research and development spending in the President's budget. However, these data are supplemented by SRS data for almost one-third (5 of 17) of its budget overview tables (AAAS 1999). In a recent report on the "new economy" prepared by the Progressive Policy Institute, five of thirty-nine indicators drew on SRS data from either *Science and Engineering Indicators* or *National Patterns of R&D Resources* compared to three drawn from *the Economic Report of the President* and seven drawn from Bureau of Labor Statistics data (Atkinson and Court 1998). Likewise, the Committee for Economic Development recently released a report, *America's Basic Research*, in which almost 90 percent of the data in the report's tables and figures were SRS data drawn from either SRS publications or the NSB's *Science and Engineering Indicators* (Committee for Economic Development 1998).

Substantial Room for Improvement

Still, national policymakers face new issues about the allocation of scarce federal resources in support of changing science and engineering. Examples of questions for which additional data may be useful follow:

- Now that the majority of scientists and engineers work outside of academia, how do they use their training in non-academic careers? Are there changes in Ph.D. training that could improve their productivity and job satisfaction in these areas?

- Which fields in which industries are generating professional opportunities for new degree holders?

- How are recent Ph.D.s faring in the job market and are they, as a group, experiencing higher levels of unemployment or underemployment than other professionals are?

- How are graduate students supported financially throughout their studies and how does the packaging of support at various stages in their training affect their success?

- How does innovation occur in a firm, and how do firms translate research findings into commercial products?

- What is the return to government-supported R&D, and how does it compare to returns on private investments?

- How has industry reshaped research and development in the private sector?

- Is the United States investing adequately in specific critical technologies?

- What forms have government-university-industry partnerships taken? Which forms have been most successful?

- What are appropriate costs associated with science and engineering facilities at major U.S. research universities?

- What are the locational arrangements among science and engineering endeavors and how are they changing at the regional, national, and international levels?

- How does the globalization of science and engineering affect the R&D enterprise and the labor market for scientists and engineers?

To provide data and analysis that will meet some of the unaddressed information needs of its customers, SRS already plans to tackle two major areas where more information is needed on emerging policy issues. First, to address issues regarding the graduate school experience, SRS is now in the planning stages for a new Longitudinal Survey of Beginning Graduate Students. SRS is initiating this survey to "improve our abilities to assess graduate education by studying the consequences of NSF support and other funding mechanisms on persistence and timeliness attributes of degree attainment, transition from education to work, and subsequent employment experiences and career outcomes." Second, in order to provide more information on the innovation process in the United States, SRS is developing a Survey of Industrial Innovation. Recognizing that "innovation and technological change play a pivotal role in the performance of U.S. firms," SRS is launching this survey "to be in the position to provide policymakers with systematic data on the processes of technological innovation" (NSF 1998a).

SRS data need to be reviewed and updated so that they better address issues such as those listed above. The three personnel surveys in the Science and Engineers Statistical Data System (SESTAT) were substantially revised in 1993, and the Survey of Industrial Research and Development (RD-1) significantly increased its sample size and added a substantial number of firms in the service sector to the sampling frame about the same time. Most SRS surveys, however, have changed little in recent years. SRS requires an ongoing process that produces, through interaction with policymakers, planners, and researchers, continual renewal of the concepts its seeks to measure and the data collection and acquisition activities that provide those measures in order to ensure the relevance of its information. A statistical agency must carry out user needs assessments on a regular basis in order to ensure that the data it provides is currently relevant. A proactive agency, however, will actively engage its constituents in a variety of ways to review and revise its data collection portfolio on an ongoing basis, maintaining important data series, yet also addressing newly emergent data needs.

Dialogue and Renewal

There has been very little systematic study on the ways statistical agencies can best engage their data users to review and update the concepts they measure. We have, however, identified a set of activities that we believe will orient SRS's processes toward continuous renewal of the concepts it seeks to measure and of its data collection and acquisition activities. First, SRS needs to organize an ongoing dialogue with its customers through advisory panels, workshops, and customer surveys. Second, SRS should obtain feedback on its data and products through outreach activities, purposeful dissemination, and enhanced customer service. Third, the analysis of its microdata—by its own staff and by external researchers—provides SRS essential opportunities for understanding its data and gathering information that will yield strategies for improving data. Fourth, SRS should create small, targeted pilot surveys on new topics using statistically valid sampling methods to determine whether they should be incorporated into the SRS portfolio. SRS should utilize the information obtained in these ways to review and revise existing surveys or enhance the data they provide through temporary survey supplements, quick response panels, or tightly-focused qualitative analyses.

Organizing a Dialogue with Data Users and Policymakers

One of the principal means for generating feedback from customers on the issues they face and the way they address them with SRS data is to establish advisory panels (called "special emphasis panels") of users, policy analysts, and data providers for each existing SRS survey. Many statistical agencies have advisory committees for their surveys, though their use varies. For example, the U.S. Census Bureau has advisory committees that focus on specific survey activities (e.g., the 2000 Census, the Survey of Income and Program Participation) as well as on a broad range of programs (e.g., the Advisory Committee of Professional Associations). By contrast, the Energy Information Agency and the Bureau of Justice Statistics each have an advisory committee for the entire agency appointed by the American Statistical Association.

For SRS, these advisory panels for SRS would seek to keep staff abreast of relevant policy issues, identify new topics for surveys, recommend survey changes, and suggest new areas for analysis. Advisory panels can also help make tradeoffs between survey questions and identify which data are practical to collect. A number of SRS surveys already have advisory panels in place. For surveys that do not, SRS should establish them. The committee is especially concerned that an advisory panel be established for the Survey of Industrial Research and Development (RD-1) to deliberate emerging issues further and to provide suggestions for revising the survey instrument in light of these issues. Also, the 1989 panel on the SRS personnel data system recommended creation of an advisory panel for what is now the SESTAT system. As SESTAT is revised for the next decade, we recommend that the Special Emphasis Panel of the Doctorate Data Project play this role, in addition to providing advice on the Survey of Earned Doctorates and the Survey of Doctorate Recipients as it does now.

Holding conferences or workshops on important science and engineering resources issues or data collection and usage issues is a second means for interacting with data users. SRS periodically holds conferences and workshops that provide useful opportunities for interchange. Examples include the series of workshops SRS has held with professional societies on issues related to education and the workforce, the 1997 NRC workshop funded by SRS on Industrial Research and Innovation Indicators for Public Policy, and the recent Workshop on Federal Research and Development that brought together agency staff who provide and use data from the

Survey of Federal Funds for R&D. These have been useful events and a set of such workshops should be planned on an ongoing basis across all SRS data collection efforts to provide interaction with data providers, data users, and policymakers and to suggest ways to keep data collection focused on data elements that address important policy issues.

SRS customer surveys obtain user judgments about data relevance and quality. In a survey of SRS customers undertaken in 1996 (with an 84 percent response rate), 38 percent of respondents rated the relevance of SRS data on the topics of most interest to them as "excellent." Another 50 percent rated them as "good." In order to mandate improvement in the relevance of SRS data, NSF has required SRS (under the Government Performance and Results Act) to increase the percentage of customers rating SRS data relevance as "excellent" to 45 percent and increase the overall percentage who rate SRS data "good" or "excellent" to 90 percent.

SRS may take other steps to improve its customer surveys and the accuracy of its results. First, SRS may better specify the universe of SRS data users it uses to draw the sample for its customer surveys. In 1996, the sampling frame for the survey was 1,200 individuals who indicated that their principal professional interest was science and technology policy in a survey conducted by the AAAS. While these individuals are likely or potential data users, they do not cover the entire universe of such users which includes federal program administrators, educators, students and faculty, academic researchers, and others, and these individuals should be included in the sampling frame. (It should be noted that a change in the sample would make the resultant data non-comparable to those derived from the 1996 customer survey, and therefore, NSF's GPRA goals for SRS of 45 percent "excellent" and 90 percent "good" to "excellent" may therefore be of questionable value.) Second, the survey may be enhanced by adding questions about the role data from specific surveys play in the work of those who

use them. In this way, customer surveys may provide SRS further insight into exactly which kinds of data its users need. An investigation of how data are used should provide an opportunity for dialogue that will improve the data for the end user.

Feedback through Outreach, Dissemination, and Customer Service

In spite of public relations and information dissemination efforts, according to one person interviewed, "SRS still isn't known well by the communities who should know it." SRS has made a significant effort to establish an accessible web site that allows interactive access to data by a wide audience. SRS also has a loyal following for data from many of its surveys, such as the Survey of Earned Doctorates, which annually generates a long list of data requests from repeat as well as one-time data users. Many of those interviewed for this study, however, felt that many people could use SRS data, but are not aware of this resource. SRS could increase its visibility by appearing at a larger number of public events, by more purposefully disseminating its data and reports, and by improving customer service.

Outreach to current and potential data user communities provides another source of interaction that informs SRS's data collection, acquisition, and analysis. SRS has engaged certain groups already. For example, the series of workshops with professional societies on graduate education and labor market issues has provided SRS an opportunity to inform users in these organizations about SRS data. The workshops allow for dialogue about current issues, and establish working relationships that help fill data gaps on such topics as underemployment for recent Ph.D.s. Similarly, SRS recently received input on the Survey of Industrial Research and Development (RD-1) from the American Economic Association Advisory Panel to the U.S. Census Bureau. While in the case of RD-1 input from an even more focused group, such as an advisory committee, is required, these linkages can be extended to other surveys. They can be augmented as well by having an SRS presence, an exhibit booth for example, at the annual meeting or conference of important constituent groups like the American Association for the Advancement of Science (AAAS), the American Economic Association (AEA), and the Council of Graduate Schools (CGS).

The data and report dissemination process provides another opportunity for meaningful interaction with customers. SRS publications appear to be distributed in an ad hoc manner. Those interested in reports contact NSF and are placed on subscription lists. SRS should examine these lists and take action to ensure that they include important constituents in the policy arena, whether or not they have requested certain publications. Some customers do not know that certain publications exist. For example, many graduate deans who regularly receive data on new doctorate degree recipients do not receive data and publications on new enrollment in graduate programs by field or on academic R&D funding that powerfully affects both graduate education and human resources needs. In addition, many SRS data users do not realize that publication of a data brief coincides with the official release of a data set. SRS should find ways to make this more transparent and to apprise its data users when data sets are made public. A short statement attached to a data brief indicating that data from a certain survey is now available to the public would help to accomplish this.

Another practical strategy for improving dissemination involves the SRS web site. A number of people interviewed for this study, including several high-volume web site users, commented that while the SRS web site is among the best hosted by a federal statistical agency, the functionality of the data sets on the web site needs to be more user-friendly.

Opportunities in Analysis to Improve Data

The analysis of its data provides SRS with additional opportunities for interacting with data users and policymakers and gathering information that will yield strategies for improving data. As is well known in the federal statistical community, a statistical agency enhances its understanding of the uses and limits of its own data by using them. Use of the data to support analyses of current issues suggests opportunities for improving existing data or augmenting them through new surveys since analysis helps to identify areas where information is incomplete and new data are needed to tackle an issue. SRS conducts a substantial amount of data analysis. SRS prepares special publications, including the influential *Science and Engineering Indicators* under the guidance of the National Science Board. SRS also prepares a series of *Issues Briefs* on current topics in science and engineering resources. Still, many SRS data sets are underutilized, to the detriment of both the user community and the process of understanding and improving SRS's data. One of our interviewees argued, "NSF has neither fully exploited its data nor sufficiently encouraged use of the data by outside analysts."

Comments by a number of other interviewees supported the claim that SRS data are underutilized and many urged that SRS institute a program that would increase use of its data by increasing access to microdata for, or providing funding to, external analysts with expertise in science and technology resources issues. Examples of the kinds of SRS data that are underutilized as perceived by interviewees include the National Survey of College Graduates (NSCG) and the Survey of Doctorate Recipients (SDR). SRS has done little comparative analysis using the available NSCG data on the non-science and engineering respondents to the survey. Also, SRS has done very little, if any, longitudinal analysis of the SDR or other surveys in its personnel system.

Increasing access to microdata is a priority item for SRS. SRS should revise its policies to increase access by researchers to microdata that are essential to analyzing the details of certain issues. Many of SRS's rich databases are not fully utilized because of the limits placed on access to detailed data due to SRS's overly restrictive interpretation of federal privacy laws. Given proper procedures, the security and confidentiality of the data may be maintained even while the number of analysts accessing microdata is increased to include a small number of external analysts.

In addition, SRS should develop a grant program to support external researchers with expertise in science and technology resources issues. We applaud the new SRS program providing young researchers training on the use of its databases. Scholarships were offered to these young researchers so that they could attend a weeklong training session. This is a step toward developing a new cohort of researchers interested in analyzing ever-complex science and engineering resources issues. Beyond this, SRS should also develop a program that provides grants to external researchers in support of analyses that utilize SRS data. SRS has the authority to administer an external grant program for analytical research, but has not issued a general announcement for one since the mid-1980s because of budget constraints. This program provided grants ranging from about $50,000 to $150,000 for external research using SRS data. Under this program, SRS requested proposals on specific topics, which provided an opportunity to obtain assessments on a variety of current issues. Moreover, according to one interviewee, these externally funded studies often anticipated issues that would otherwise have taken two to three additional years to surface. In this sense, these grants not only inform current issues but also provide what some have termed "exploratory indicators." In essence, these studies used SRS data to help identify emerging issues and to define the kinds of data that will be needed to address these issues. We would like to see a new grants program continue this.

SRS should also develop a means for bringing researchers into SRS to work on science resources issues, perhaps for six to twenty-four months. The development and funding of these programs and the resultant familiarization of these analysts with SRS data will strengthen SRS's programs substantively and methodologically. Independent external researchers may explore science and engineering resources issues in new ways or tackle issues that SRS staff may not have the time to explore. This research would broaden the range of insights developed from SRS data. Moreover, independent researchers—as do SRS analysts—would uncover methodological or data quality issues in SRS data sets, thereby helping to improve the data as well. A variety of issues needs to be addressed in developing programs such as these. For example, should a researcher who comes to work at the SRS study issues defined by SRS, independently defined issues, or some mixture of the two? Would analyses result in SRS publications or independent publications? Certainly details such as these will have to be addressed, but the experiences of these researchers, even if few in number, would contribute to SRS's mission by increasing the flow of ideas in and out of the organization. NSF has standards and procedures for bringing visiting personnel into other programs that might be adapted for SRS. Statistical agencies like the Census Bureau also have regular visitor programs that could serve as models.

Dissemination of analyses provides an opportunity for additional feedback from data users and policymakers. It is difficult for a statistical agency to know *a priori* all of the issues that matter to its customers since important issues are ever changing. Analyzing data on subjects that are known to be important allows for additional opportunity to interact with data users and policymakers to generate additional feedback that may identify new issues for analysis or new data products that serve customer needs.

Using Feedback to Revise Data Collection and Acquisition

SRS should approach each of its data collection efforts as an opportunity to provide important time-series data as well as data on current issues. Many analysts find time series valuable, and SRS should collect data so as to allow time series analysis in each of its substantive areas. Structural changes in the science and engineering enterprise, however, also require changes from time-to-time in the questions posed and response categories offered in data collection efforts. The feedback generated through dialogue, outreach, dissemination, and analysis should be used to address these structural changes by revising—and occasionally overhauling—data collection activities to meet current data needs.

SRS should establish a periodic review process for each survey to ensure that they address the current structure of the science and engineering enterprise based on the feedback obtained through the means described above. With guidance from survey advisory committees, staff responsible for each survey should be required to periodically submit a plan to the survey's advisory committee for dropping obsolete questions and adding new ones as needed to keep the data collection effort current.

To tackle new dimensions of the science and engineering enterprise that may substantially change data, SRS has to consider the difficult tradeoffs between maintaining time series and changing data to reflect new realities. The Committee on National Statistics has said that a federal statistical agency must "be alert to changes in the economy or in society that may call for changes in concepts or methods used in particular data sets. Often the need for change conflicts with the need for comparability with past data series and this issue may dominate consideration of proposals for change" (NRC 1992).

The assumption should be that it is better to maintain than disrupt a time series, but when reality has changed substantially or it is determined that current approaches to data collection do not capture the sources and uses of science and engineering resources, it is better to start a new time series that provides relevant information. Changes to the personnel surveys in the SESTAT system and to RD-1 earlier in the 1990s, for example, did result in discontinuities in time series. However, the need to better expand and integrate survey instruments, in the case of the three personnel surveys, and the need to improve sample design, in the case of RD-1, compelled a break with the past in these instances. To strike the appropriate balance between maintaining and disrupting time series in any given situation, SRS should draw on the advice of its data users through its survey advisory committees to establish what action would best meet their needs.

When structural changes like these are not the issue, there are ways to address current issues without disrupting time series. First, to address important issues that may not require permanent changes in a survey instrument, SRS should continue to use special survey supplements to obtain needed data. Some SRS surveys already use special supplements. For example, the 1995 Survey of Doctorate Recipients (SDR) included a work history module. Other surveys, such as the Survey of Earned Doctorates (SED), have never used such modules even though they could be employed profitably to obtain data that policymakers desire. Second, SRS may respond even more quickly to important short-term issues by using quick-response panels to answer questions that require a quick turnaround rather than changing an existing survey instrument. SRS had an Industrial Panel on Science and Technology and a Higher Education Panel that it used in the 1970s and 1980s to answer questions quickly when needed and without substantial cost.

SRS often used these panels, for example, to gather time-sensitive information in response to questions posed by the U.S. Congress and OMB. Examples of the issues previously examined through quick-response panels include the impact of the R&D tax credit and foreign support of research conducted on U.S. campuses.

SRS should also consider qualitative techniques in expanding its repertoire of methods for information collection. Quantitative data cannot answer every question, or at least, not in the most cost-effective manner. Just as SRS can use quick-response panels to explore an issue broadly, it could profitably use case studies as a means for exploring particular issues that require in-depth investigation or for which data are not obtainable. Case studies would be appropriate when understanding an issue or phenomenon requires more detail than a survey could cost-effectively provide. SRS could also use focus groups, site visits, and interviews as means for collecting information. SRS already uses focus groups as a mechanism for pretesting new questions and response categories on its surveys. However, these techniques could also be used to collect primary information as part of a quick-response activity or a case study.

Data Linkages

The second aspect of data relevance that federal statistical agencies should focus on is the ability to link their data sets internally and to data collected by other sources. The Committee on National Statistics asserts that an effective statistical agency promotes data linkages in order to enhance the value of data sets for analyzing current policy, program, and research issues (NRC 1992). Science and technology policy analysts and researchers interviewed for this study also expressed a desire for more cross-survey analysis to provide better interpretations of the changing components of the science and engineering

enterprise and how they interrelate. They have suggested that integrating data sets within SRS and better coordination of SRS data collection with that of other organizations will facilitate these kinds of analyses.

Integrating SRS Data Sets

SRS's portfolio of data collection activities has grown over the past half century as individual surveys have been established to provide information on specific pieces of the science and engineering enterprise. SRS has only recently begun to manage its surveys as components of a more integrated data system and would increase the depth and validity of its data by pursuing this further. A number of interviewees for this study claimed there are many "walls" in SRS among staff and between surveys, a suggestion borne out by differences among survey instruments and in survey results. SRS should facilitate greater interaction among staff to promote information sharing, to examine how SRS surveys could be better coordinated, to seek practical means for linking data sets, and to surface opportunities for cross-survey analysis.

WebCASPAR and SESTAT, discussed in the previous chapter, provide examples of ways in which SRS has integrated its data and could do so in the future. WebCASPAR draws on data from SRS and the National Center for Education Statistics to provide profiles of academic institutions. SESTAT, which will be discussed further in Chapter 4, combines data from the National Survey of College Graduates, the Survey of Recent College Graduates, and the Survey of Doctorate Recipients to provide an integrated data system on the population of scientists and engineers at the bachelor level and above in the United States.

On the whole, however, SRS surveys have tended to operate relatively independently of each other. SRS could implement changes to make data elements more consistent across surveys. SRS should find ways to link data sets beyond SESTAT and WebCASPAR to facilitate deeper analysis of the rich data sets it has created and maintained.

First, SRS should improve comparability of data derived from its surveys within each of its statistical programs. For example, the Survey of Earned Doctorates and the Survey of Doctorate Recipients both ask Ph.D. recipients questions on such subjects as type of employer and disability status. However, the two surveys ask these questions in different ways and with different response categories. As the two national surveys of Ph.D.s, the content of these two surveys could be better coordinated to link the status of Ph.D.s at degree receipt—the point at which the SED is administered—with the career paths of Ph.D.s, the subject of the SDR.

Second, SRS should continue its investigations of discrepancies in the results it obtains on similar questions from different surveys in order to assess and improve the validity of its data. SRS has been investigating several recently identified data discrepancies in its R&D statistics program. For example, the Survey of Federal Funds for Research and Development reports federal obligations of $31.4 billion in 1997 for research and development performed by industry, while the Survey of Industrial Research and Development reports that the federal government was the source of just $23.9 billion in industry R&D expenditures (NSF 1999b, NSF 1999h). Similarly, the Survey of Federal Funds for Research and Development reports federal obligations of $12.6 billion in 1997 for research and development performed by the nation's colleges and universities, while the Survey of Research and Development Expenditures at Universities and Colleges reports that the federal government was the source of $14.5 billion in academic R&D expenditures (NSF 1999b, NSF 1999h). This discrepancy and some of its analytical consequences will be discussed further in Chapter 5.

In the 18 months or so that these discrepancies have been evident, SRS has undertaken a number of activities (e.g., workshops with data providers in federal agencies and academic institutions) to understand the nature of these discrepancies, the extent to which some differences might be expected and can be explained, and the extent to which some differences need to be addressed through changes in either questionnaires or data collection.

SRS has noted that the disparity in the results of the Survey of Federal Funds for Research and the Survey of Industrial Research and Development can be accounted for by a discrepancy between what the Department of Defense (DOD) reports as the R&D funding it provides to industry and what industry reports as the R&D funding it receives from DOD. Thus, SRS has been able to suggest what some of the reasons for the disparity may be: firms may not count as their own R&D those projects funded by federal agencies that they subcontract to non-industrial organizations; DOD may classify some technical services contracts with industry as R&D that firms do not classify as such; there are differences between federal obligations and actual expenditures from year-to-year, though these should be small; and there may be differences between what is reported as R&D by a firm as opposed to what might more accurately be reported as R&D by a firm's business unit. SRS has discovered that other advanced industrial countries with significant defense programs experience similar accounting problems.

SRS has begun an investigation into the discrepancy between the Federal Funds Survey and the Survey of Research and Development Expenditures at Universities and Colleges. SRS has found that at least $350 million of the $1.9 billion difference—almost one-fifth—is due to double counting that occurs when one university subcontracts R&D to another and they both count it as R&D expenditures.

Another potential source of the difference may occur when federal agencies obligate funds to states or industry, which then pass funds on to universities. The federal government counts these funds as obligated to states or industry, as appropriate, and not as R&D funding to universities. The universities, however, knowing the original source of these funds, report them as federally-funded R&D expenditures.

SRS should continue these investigations and take steps in the near term to explain these differences to data users and consider implementing questionnaire or data collection changes that will help resolve the discrepancies. We additionally recommend that SRS consider commissioning a further study that would examine in greater detail the possibility of better integrating the SRS R&D surveys. Such a study might design an "R&D data system" analogous to the SESTAT data system in the Human Resources Statistics program. The resulting data system would ideally account for related structural changes in the nation's research and development system, such as intra- and inter-sectoral R&D partnerships.

Third, SRS should structure its surveys so they may also be directly linked one to another. There is great potential utility in such linkages, including those between human resources and R&D investment data sets. The most critical obstacle to meaningful linkage of SRS data sets, other than non-comparability in questions and response categories, has been the lack of a uniform science and engineering field taxonomy across SRS surveys. As can be seen in Table 3-1, for example, the biological sciences are described differently across SRS graduate education and personnel surveys. Each of the SED, the GSPSE, and the NSCG use a different taxonomy for science and engineering field of degree. To cite another

example, there are differences in field classifications among R&D funding and performance surveys. The life sciences in the Survey of Scientific and Engineering Expenditures at Universities and Colleges are disaggregated into agricultural, biological, medical and other life sciences. In the Survey of Federal Support to Universities, Colleges, and Non-Profit Institutions, however, the life sciences are disaggregated into these fields as well as a fifth, environmental biology. SRS has contracted with SRI International to review its taxonomy. The SRI analysis should be taken seriously as a step toward standardizing taxonomies across data collection activities.

Field taxonomies not only require standardization so that different surveys may be better linked; they also require a means for keeping them current on an ongoing basis so that data may continue to be relevant to policymakers. It has been known for some time that SRS taxonomies become outdated and need constant revision. An internal 1994 SRS report on customer views of SRS products stated:

> Many customers, especially those at NSF, said that some science and engineering taxonomies used in SRS surveys were out-dated and/or not at the level of detail necessary to provide the information they need (NSF 1994).

Care should be taken to preserve time series when possible, but taxonomic changes are required in order to ensure that the data are providing accurate measures of current science and engineering R&D funding and performance, especially in emerging fields.

Table 3-1 Biological Science Fine Field Categories in the 1997 Survey of Earned Doctorates (SED), the 1997 Survey of Graduate Students and Postdoctorates in Science and Engineering (GSPSE), and the 1997 National Survey of College Graduates (NSCG).

SED Biological Science Fields	GSPSE Biological Science Fields	NSCG Biological Science Fields
Biochemistry	Biochemistry	Biochemistry and Biophysics
Biomedical Sciences		
Biophysics	Biophysics	
Biotechnology Research		
Bacteriology		
Plant Genetics	Botany	Botany
Plant Pathology		
Plant Physiology		
Botany, Other		
Anatomy	Anatomy	
Biometrics & Biostatistics	Biometry and Epidemiology	
Cell Biology	Cell and Molecular Biology	Cell and Molecular Biology
Ecology	Ecology	Ecology
Developmental Biology/ Embryology		
Endocrinology		
Entomology	Entomology and Parasitology	
Biological Immunology		
Molecular Biology		
Microbiology	Microbiology, Immunology, and Virology	Microbiology
Neuroscience		
Nutritional Sciences	Nutrition	Nutritional Sciences
Parasitology		
Toxicology		
Genetics, Human & Animal	Genetics	Genetics, Animal and Plant
Pathology, Human & Animal	Pathology	
Pharmacology, Human & Animal	Pharmacology	Pharmacology, Human & Animal
Physiology, Human & Animal	Physiology	Physiology, Human & Animal
Zoology, other	Zoology	Zoology, General
Biological Sciences, General	Biology, General	Biology, General
Biological Sciences, Other	Biology, not elsewhere classified	Other Biological Sciences

49

Coordination and Linking with External Agencies and Organizations

While it is a high priority for SRS to focus on better data integration among its own surveys, there are opportunities for SRS to improve the coordination of, and create linkages between, its survey data and those of other federal agencies and non-governmental organizations.

First, SRS could help coordinate data collection by professional associations, colleges and universities, and other countries to improve data availability and comparability.

- As will be further discussed in Chapter 4, SRS needs to continue to improve its data on the transition of new Ph.D.s to employment. SRS should continue its productive work with others to obtain data on the job market experience of new Ph.D.s—especially those without definite commitments for study or work at the time of degree—during the immediate months following graduation. SRS has interacted extensively and productively with professional societies and the Commission on Professionals in Science and Technology (CPST) in obtaining data on the job market experiences of new Ph.Ds. This interaction should be continued, and strengthened. SRS could provide these associations with a framework for collecting these data in a standardized way and technical assistance on statistical methods.

- SRS should also explore how it might work closely with colleges and universities to assist them with the development of standardized data sets on the placement of their recent graduates. In its recent report on graduate education, the Association of American Universities (AAU) urged colleges and universities to maintain comprehensive data on completion rates, time-to-degree, and job placement for each of their graduate programs. The report specifically recommends that institutions track their graduates at least until first professional employment beyond postdoctoral appointments. Such data would allow better tracking of alumni for purposes of understanding career paths and the effectiveness of university programs in providing the knowledge and skills that graduates need in the labor market. The AAU also suggests that institutions provide student applicants with this information (AAU 1998). As research universities implement this recommendation, SRS could help them standardize their data collection efforts; locally collected data could then potentially be aggregated in a meaningful way at the national level.

Second, while meeting requirements for maintaining confidentiality, SRS should seek to link its data with those of other federal agencies and private firms to further enhance their analytic value:

- SRS should seek ways to link its graduate education and personnel data with data collected by the other relevant data sources such as the National Center for Education Statistics and the Educational Testing Service (ETS). For example, researchers would like to be able to link student scores on the Graduate Record Examination (GRE) to data from the Survey of Earned Doctorates and the Survey of Doctorate Recipients to examine further the predictive power of the GRE with regard to career outcomes.

- SRS data could be linked to an array of other education, career, and productivity data. SRS should find efficient means for linking its graduate education, personnel, and R&D funding and performance data with public data on federal agency awards (fellowships, traineeships, associateships, research grants) and patents and with private data sets on publications. For example, graduate students can accurately identify the type but not the source of their funding, particularly when funding originates from federal agencies and flows through institutions. Creating linkages between federal data sets on awards and federal data sets on degree recipients would substantially improve the available data on federal support for graduate students. In theory, linking SRS data to other federal data sets, particularly those that include name and Social Security Number as identifiers, should be routine, though in some cases augmenting data sets on awards will need to be improved to facilitate the linking process. Linking personnel data with publication and patent data is more difficult, but SRS should also seek ways to make such linkages easier. SRS asked for data on the number of articles, other publications, patents applications, and patent awards in a special module for the 1995 NSCG and 1995 SDR. If these data points were collected on a continuous basis they would provide benchmarks against which researchers could measure their ability to link SRS with publication and patent databases.

- SRS should also investigate the opportunities made available by adding an indicator for metropolitan statistical area to each of its R&D funding and performance surveys. Currently, SRS R&D surveys collect data on the state in which the respondent is located which provides an approximation of the state distribution of R&D performance. (Actual location of performance may differ from location of respondent when laboratories or business units are geographically dispersed.) These data are widely requested by federal and state officials and by researchers who are interested in the locational dimensions of R&D spending.

However, the data on state distribution of R&D spending does not offer a precise measure of the regional economic impact of R&D expenditures since regional economies often include areas in more than one state and there may be more than one regional economy within a state. These data users, particularly researchers, would benefit from a metropolitan indicator as an added data point for analysis and also as a means for linking these data with other economic and demographic data available at the metropolitan level.

Finally, SRS interprets federal privacy laws in a way that restricts the agency's ability to provide data to external researchers and specifically to facilitate data linkages. SRS does provide data to external researchers under licensing agreements. However, we believe that SRS is overly restrictive in its interpretation of privacy laws, and thus limits both the use and usefulness of the data it provides to these researchers. With proper licensing agreements and oversight, SRS should be able to better support the data linking activities of its data users.

Data Currency

A third dimension of relevance for statistical agencies relates to whether data collected by such agencies are current, or are perceived to be, during their "shelf life," the period of time from the date they are made public until the date when data from the next survey cycle are released. Using the Survey of Doctorate Recipients (SDR) as an example, Figure 3-1 illustrates how survey periodicity and data processing time affect data currency. The survey reference date for the 1995 SDR was April 15, 1995. Surveys were sent to respondents in May 1995, data were collected throughout 1995 and early 1996, and file preparation was completed toward the end of

Figure 3-1 Survey Reference Date and Data Release Date for the 1995 and 1997 Cycles of the Survey of Doctorate Recipients

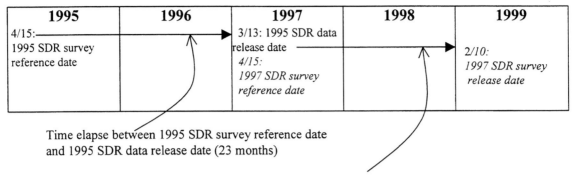

1995	1996	1997	1998	1999
4/15: 1995 SDR survey reference date		3/13: 1995 SDR data release date *4/15: 1997 SDR survey reference date*		*2/10: 1997 SDR survey release date*

Time elapse between 1995 SDR survey reference date and 1995 SDR data release date (23 months)

Expected "shelf life" of the 1995 SDR data between 1995 SDR data release and 1997 SDR data release (23 months)

1996. NSF released the data set on March 13, 1997. Thus, approximately 23 months elapsed between the survey reference date and the data release date. Given the two-year periodicity of the survey, then, the 1995 data were the most recent data available on the population of Ph.D. scientists and engineers until February 1999, when NSF released the 1997 SDR data by posting a limited set of early release tables on their web site.

At issue is whether the 1995 SDR data were accepted by their audience as "current" for the entire length of their expected "shelf life," i.e., the 23 months from March 1997 to February 1999. Experiences of SDR analysts suggest that by mid-1998, some already questioned the validity of analyses based on 1995 SDR data. For example, when the NRC's *Trends in the Early Careers of Life Scientists* (NRC 1998c) was released, its detractors argued that the labor market for scientists and engineers had already changed and 1995 data used in the analysis did not reflect these changes. Even if, in fact, the situation in 1998 was much the same as in 1995, the lack of more recent data allowed those who wanted to substitute anecdote for data in policy debates to do just that. To make matters worse, the 1997 SDR had already been administered, but its data had not yet been

released. Thus, the period of data currency did not extend throughout the expected shelf life of the data.

In the case of the SDR, a shorter time period between survey cycles could increase the percentage of its expected shelf life that its data are current. However, shortening periodicity increases survey costs, potentially doubling them if the periodicity is cut in half from a biennial to an annual cycle. By contrast, SRS has considered increasing the periodicity of the SDR from two to three years to cut costs. One might argue that the labor market for Ph.D. scientists and engineers does not change substantially in two years and that three years is sufficient to capture important trends. Such a periodicity change would save money, then, while not affecting data accuracy. However, such a move would substantially decrease the percentage of the shelf life period that SDR data are current, and therefore, would reduce their overall relevance and usefulness to analysts.

Data currency also depends on how long it takes SRS to administer a survey (including survey follow-up), process survey data, and release them for public use. This process is measured, in the case of SRS, as the time that elapses between the reference date in each survey and the date at which survey data are released. The timeliness with which SRS releases its data has been of comment for some time. A 1994 customer survey conducted by SRS concluded that "timeliness of SRS data was a major concern of several interviewees; it was not considered a problem by others." In summarizing the findings of a 1996 customer survey, then-SRS deputy director Alan Tupek concluded, "when interest in a topic is high, currency/timeliness is the major area of concern. I conclude that we should continue to place priority on improving timeliness" (NSF 1997). Timeliness has also been a concern of external analysts. The Committee on Science, Engineering, and Public Policy (COSEPUP) of the National Academies found in utilizing data from the Survey of Earned Doctorates and the Survey of Doctorate Recipients that the data were not sufficiently current to provide timely measures of the rapidly changing environment for the education and labor market for scientists and engineers. COSEPUP concluded "The National Science Foundation should continue to improve the coverage, timeliness, and clarity of analysis of the data on the education and employment of scientists and engineers in order to support better national decision-making about human resources in science and technology" (NAS 1995).

The relevance, and therefore, the use of SRS data will increase as the data are delivered in a more timely fashion. The *NSF GPRA Performance Plan for FY1999* requires SRS to improve its timeliness by decreasing by 10 percent from the current average of 540 days, the time interval between the reference period (the time to which the data refer) and the data release date for each survey. Table 3-2 provides information on SRS's record for data releases. There are several options for decreasing the time between reference and release dates. Some options focus on the data collection phase; other options focus on the data processing and analysis phase.

- **Reduce the time for data collection by tightening the schedule for the initial**

and follow-up phases of survey response. At present, many SRS surveys are conducted by mail with computer-assisted telephone follow-up. Minimizing the time for each of these steps, consonant with maintaining adequate response rates, could potentially save weeks or months in the schedule. As an example, the methodological report for the 1995 Survey of Doctorate Recipients states that a paper questionnaire was sent to sample members in May 1995, a second questionnaire was sent by priority mail in July, and computer-assisted telephone interviewing (CATI) was used for follow-up between October 1995 and February 1996. It seems worthwhile to explore the extent to which the data collection period for the SDR could be reduced from 9 months to 6 months or even less. Analysis of mail-back rates by week would be helpful in this regard. If, for example, the response to the first mail questionnaire dropped off substantially after 3 or 4 weeks, it would make sense to mail the second questionnaire in June rather than July. Similarly, it might be possible to begin the CATI operation as early as August and complete it within 2 or 3 months.

- **Explore further the use of Internet technology as a survey mechanism that could speed up response**. It would be useful to conduct research to determine if respondents to such surveys as the SDR would respond more willingly and faster via the Internet.

- **Explore the use of motivational material in the survey mailings that emphasizes not only the need for a high response rate, but also the need for a timely response**. In addition, for longitudinal surveys, such as the SDR, or the large businesses that are regularly included in RD-1, early results might be mailed to respondents to thank them for their timely response.

Table 3-2 Time Elapse in Months for SRS Surveys Between Reference Month and Data Release Month (data displayed by calendar year of reference month) (in months)

	1993	1994	1995	1996	1997
National Survey of College Graduates*	25	n/a	25	n/a	16
National Survey of Recent College Graduates*	21	n/a	19	n/a	21
Survey of Doctorate Recipients*	23	n/a	23	n/a	22
Survey of Graduate Students and Postdoctorates in Science and Engineering	18	18	16	16	13
Survey of Earned Doctorates	16	17	11	17	18
Survey of Scientific and Engineering R&D Facilities at Colleges and Universities*	--	n/a	12	n/a	17
Survey of Research and Development in Industry	19	17	23	15	13
Survey of Research and Development Expenditures at Universities and Colleges	15	16	14	15	13
Survey of Federal Support to Universities, Colleges, and Nonprofit Institutions	20	21	20	20	16
Survey of Federal Funds for Research and Development	20	21	20	19	14

*Biennial surveys: n/a means not applicable

- **Release data for key indicators early**. For many key business statistics (e.g., gross domestic product figures) preliminary results are released as soon as possible followed by revisions at a later date. Such a strategy might be explored for some SRS statistics. The preliminary results could be based on early survey responses, on data for a sub-sample or on simpler data editing and imputation procedures than would ultimately be used for the final estimates. Alternatively, final estimates for key indicators could be released early by putting the processing, analysis, and data review for them on a fast track separate from the other data collected in a survey. (As an example, the monthly unemployment figures from the Current Population Survey are released within a few weeks of data collection, while other data in the survey, such as income statistics, are released on a slower schedule.) Central to this strategy is the identification of statistics from a survey that are most important for key users. Strategies to permit their early processing and release then need to be devised. Such strategies may include an emphasis on obtaining the responses for these indicators in the follow-up of non-respondents.

Research will be required to determine which of these or other options is feasible and cost-effective for each of the SRS surveys. Given the interest in timely data for many important policy issues related to science and engineering resources such research is a high priority. In addition to the options outlined above, quick-response surveys are another way to provide timely data on topics of current interest that may not yet be covered in existing surveys.

Part II

Relevant Data for Issues in
Science and Engineering Resources

The previous chapter described how federal statistical agencies should ensure the relevance of the information they provide by monitoring changing policy issues and concepts, identifying useful data linkages, and improving data currency. The following two chapters outline current issues in the changing science and engineering enterprise about which SRS could improve the relevance of the data and analyses it provides to policymakers, researchers, and others in order to better inform the decisions they make.

The analysis here highlights the importance of specific science and engineering resources issues that warrant attention by SRS at this time. We identified these issues through briefings given by SRS managers, interviews and focus groups with SRS staff and data users, and a national workshop on measuring resources for science and engineering. Committee members discussed these issues and separated them into those that SRS is already addressing and those that require further changes or investigation. We then sought to

substantiate that these latter issues are of sufficient consequence to warrant further attention by SRS. First, we examined the recommendations of national reports on science and technology policy, graduate education, and the labor market for scientists and engineers. Often, these recommendations not only substantiated the critical nature of certain issues, but also made specific observations about the data needed to address them. Second, we examined data trends, where possible, to draw our own conclusions about the direction of change and the importance of certain issues. SRS did not provide the committee with complete data on its budget and how it is distributed across staff salaries and benefits, survey operations, data analysis, publications, and other activities. Our analysis, therefore, does not attempt to provide a cost-benefit analysis that would have led to further prioritization of changes in SRS surveys that we suggest or to a protocol for trade-offs among current and proposed activities. In a limited number of cases we do raise the issue of cost and suggest that the costs of obtaining certain data should be weighed against the benefit of collecting them.

As we identified issues of current concern to the science and engineering community, we were aware that, given the short time frame and limited resources within which we were working, it would be impossible to be exhaustive in our review of the SRS data collection portfolio. Therefore, we describe what we believe are important examples of issues that SRS should better address. Others may well identify additional issues. Also, we describe these issues in a general way. It is our hope that advisory committees for each SRS survey will continue the work of this committee by further investigating current and emerging issues and by providing SRS staff with recommendations for operationalizing changes in specific survey instruments to better measure the new concepts that have been identified.

We should also point out that, in examining data collection activities across the SRS portfolio, we identified only a few items that we believe should be deleted. We concur with the suspension of the Survey of Scientific Instruments and Instrumentation Needs. We believe that SRS's resources are better spent on other activities and issues. We also specify that the collection of data in the Survey of Industrial Research and Development on the product class of applied research and development be dropped, but we also suggest that SRS examine the costs and benefits of fielding the survey at the line-of-business level rather than at the firm level. We believe that other SRS activities should remain in the division's portfolio. Again, specific items within these activities may be dropped as necessary when identified in the future by advisory committees or through other forms of dialogue with the data user community.

4

Science and Engineering
Human Resources

Graduate Education of Scientists and Engineers

Science and engineering graduate education in the United States is widely regarded as the leading system for advanced training in the world (Bowen and Rudenstine 1992, NRC 1995, AAU 1995). Several factors contribute to its quality and effectiveness. The decentralized nature of graduate education in the United States is one of its defining characteristics. More than 600 institutions provide graduate education at the masters or doctorate level in these fields, and within these institutions a range of departments and their faculty oversees the training of graduate students. The integration of graduate training with scientific and technological research is another hallmark of graduate education in the United States, especially at the doctoral level. Doctoral candidates in science and engineering typically work on advanced research projects as research assistants and also complete a research-based dissertation. Federal support of graduate students, particularly through their participation in research activities, is a third prominent feature

of advanced science and engineering education. "A major objective of the federal/university partnership in research and education historically," the National Science Board (NSB) has written, "has been to attract high-ability youth into science and engineering careers by providing significant multi-year financial support that is competitively allocated and based on the student's past achievement and future promise" (NSB 1997). Much of the federal funding for graduate students is a by-product of federal support for academic science and engineering research, but it nevertheless remains substantial and has served as the vehicle for integrating graduate education and research in science and engineering. As seen in Table 4-1, the federal government was the primary source of support for 20 percent of graduate students in 1997. The federal government provided the primary source of support for almost half of the graduate students with research assistantships and for more than half of those with traineeships. Other important sources of support were institutional support for teaching assistantships and self-support (own, spouse, or family resources) (NSF 1999e, NSF 1999l).

Table 4-1 Full-Time Graduate Students in Science and Engineering Fields, by Mechanism and Source of Financial Support, 1997

Source of Support	Fellowships	Traineeships	Research Assistant-ships	Teaching Assistant-ships	Other	All
Federal	6,255	5,155	41,370	776	3,133	56,689
Institutional	15,233	3,877	29,311	60,693	10,163	119,277
Other	4,968	635	12,526	583	4,480	23,192
Self-Support	0	0	0	0	81,454	81,454
All	26,456	9,667	83,207	62,052	99,230	280,612

Source: NSF/SRS, Survey of Graduate Students and Postdoctorates in Science and Engineering (1999l).

Continuing excellence in graduate education and the preparation of our nation's scientists and engineers for research remains an important objective for universities and the federal government, but the context for graduate education and the nature of science and engineering careers have changed in recent years. A series of reports has raised questions about the role of the federal government in providing financial support for graduate students, the progress of students through graduate school, the content of graduate education, and the preparation of graduate students for science and engineering careers. These reports have also raised questions about the ability of existing data to address these questions.

Reports on Graduate Education

In 1995, the National Academies' Committee on Science, Engineering, and Public Policy (COSEPUP) published *Reshaping the Graduate Education of Scientists and Engineers*. Citing changes in the job market for Ph.D.s, COSEPUP called for changes in graduate education and the way graduate students are supported to better position graduates for the job market and their future careers (NAS 1995). In the wake of the COSEPUP report, the National Science Board also examined graduate and postdoctoral education in the context of the federal-university research partnership. The NSB recommended that mission agencies recognize "the intimate connection between research and graduate education in universities" and adopt practices to ensure that their funding "reaps the dual benefits of simultaneously advancing both research and graduate education." The Board also recommended that federal agencies "recognize and reward institutions that, in addition to the core Ph.D. education, provide a range of educational and training options to graduate students...tailored to the career interests of the individual Ph.D. candidate" (NSB 1997). In 1998, the Association of American Universities (AAU) released a report on graduate education that noted the common criticisms of graduate education in the 1990s: overproduction of Ph.D.s; narrow training; emphasis on research over teaching; use of students to meet institutional needs at the expense of sound education; and insufficient mentoring, career advising, and job placement assistance. However, this report reached different conclusions about the nature and extent of these problems, and in the end, different recommendations (AAU 1998).

While they differed in diagnosis and recommendations, each of the NSB, COSEPUP, and AAU reports recommended collecting additional data on, and conducting additional analyses of, graduate education and the careers of scientists and engineers. The NSB report asked for "improved policy data to assess the effectiveness of current Federal support for graduate education including attention to attrition and time-to-degree." The NSB also asked for improved policy data "to identify current and emerging national needs for the science and engineering workforce" (NSB 1997). COSEPUP asked the National Science Foundation to collect improved data on "time to employment" (including unemployment and underemployment) of recent Ph.D.s and on the academic and nonacademic careers of Ph.D.s (NAS 1995). The AAU recommended that universities improve tracking of their graduates for up to five years in order to generate better information on Ph.D. placement and employment (AAU 1998).

Obtaining Improved Data on Graduate Education

SRS is responsible for collecting and acquiring key data relating to the policy issues in graduate education and the labor market for scientists and engineers that the NSB, COSEPUP, and the AAU have identified. To improve the data it provides policymakers and others on graduate education, SRS needs to explore ways to collect additional data in a cost-effective manner to fill several specific gaps in information that policymakers need.

SRS may collect data on graduate students, their educational experiences, and their financial support at three points in time: during graduate school, at the completion of a graduate degree, and retrospectively through personnel surveys. Currently, SRS administers a Survey of Graduate Students and Postdoctorates in Science and Engineering

(GSPSE) that collects data on current graduate students and postdoctorates through their institutions. SRS collects data on the universe of new Ph.D.s in science and engineering through the Survey of Earned Doctorates at the time the Ph.D. is received. SRS has also recently collected limited retrospective data on the graduate school experience through a one-time module added to the 1997 Survey of Doctorate Recipients.

To expand the range of data collected on graduate education, SRS is exploring the potential for fielding a longitudinal survey of beginning graduate students. This survey would be designed to collect data on a panel of students from matriculation through degree attainment or attrition in order to fill gaps in the data on graduate education such as those identified above. Questions that solicit data on the following could potentially be considered for such a survey:

- Financial support--longitudinal tracking of types and level of support by year
- Graduate school experience--decision to enter graduate school and career expectations; reasons for attrition (if applicable), and relationship with mentor and department
- Time to degree--data on educational milestones
- Graduate school training--responses to the depth and breadth of knowledge gained

The overall goal of a survey of graduate students would be to inform decisions made by the NSB and others concerned with graduate education, and in particular, how types of financial support affect outcomes. It would also provide data to graduate students and mentors on the graduate school experience.

SRS should carefully examine the potential for developing and administering such a longitudinal survey of beginning graduate students, yet it should defer implementation of such a survey until additional information on graduate education is gathered and analyzed. Based on what we currently understand, we question whether such a survey would be able to obtain useful data in a cost-effective manner. Financial support packages that graduate students receive during their course of study may be difficult to define because students do not always know the source of their financial support and administrators rarely have a complete picture of student financial support. Even if respondents were able to provide valid responses to questions on this subject, it is not clear that the data would be useful unless also linked to data on both the academic standing of students at matriculation, e.g., grade-point averages and Graduate Record Examination (GRE) Scores and later career outcomes. Having all of these data would greatly improve our understanding of graduate education, but we question whether the benefit of collecting this information longitudinally for a large enough sample would be worth the very large expense it would require.

We recommend that in exploring whether to develop a survey in this area SRS should assess whether quality data could be obtained and at what cost by taking these steps:

- thoroughly review the literature on the research undertaken on the graduate school experience, particularly on financing graduate education, and determine which issues have not been resolved
- address why these issues have not been resolved
- delineate the data that are necessary to address these issues
- establish whether these data can be obtained and, if so, from whom
- determine the costs associated with obtaining the data that could be collected.

If SRS determines that a survey of graduate students could be successful and cost-effective, it should be longitudinal in

nature, but only if NSF intends to support and utilize it as such. Longitudinal surveys are very rich data sources. Also, questions asked over time in a longitudinal format will have only minimal response errors associated with memory. For example, the same person may respond to the query "Why did you go to grad school" very differently at time of matriculation as opposed to retrospectively at time of exit. However, SRS has done very little to exploit the longitudinal nature of the personnel surveys they already administer.

Financial Support for Graduate Students

The National Science Board asked in 1997 for an exploration of "improved policy data to assess the effectiveness of current Federal support for graduate education" (NSB 1997). The Survey of Graduate Students and Postdoctorates in Science and Engineering (GSPSE) and the Survey of Earned Doctorates (SED) provide data on the financial support of graduate students, but neither provides data that present a sufficiently complete picture of how students are supported over the course of graduate school. GSPSE provides institutional counts of students by financial support category, but does not track individual students. The SED asks respondents for type of support (fellowships, traineeship, teaching or research assistantship, etc.), but its data do not provide reliable information about the source of support (university, federal agency, foundation, etc.). Students have trouble identifying the original source of their funding, especially when federal funds pass through their institution (NRC 1994). The SED also asks for primary and secondary type of support, but the terms "primary" and "secondary" are insufficiently described to be clear as to whether respondents are thinking "largest" or "most important" when they answer this question. Moreover, SED data do not provide a picture of financial support for an individual by year (or smaller unit of time) over the course of graduate school. It only asks respondents to indicate what kinds of financial support they had during graduate school in the aggregate. Finally, SED data are available only for those individuals who complete their course of study.

The ability to relate different types of support—especially "packages" of support over the time spent in graduate school—to the graduate school experience, graduate training, and short- and long-term career outcomes, would provide policymakers with means for evaluating effects of types of support (research assistantships, fellowships, and traineeships) that federal agencies, such as NSF, offer to students. To best analyze the efficacy of certain types of support provided to students, policymakers could utilize data that show how students were supported at different points as they proceeded through graduate school and how these patterns of funding affected students' education, progress, degree completion, and job outcomes. They would also need to have an indicator of student potential in graduate school, such as undergraduate grade point average (GPA), selectivity of bachelor's institution, or GRE scores, and indicators of career outcomes so that student potential, graduate financial support, and career outcomes could be analyzed as a package. If it were possible and cost-effective to obtain data on student potential and graduate school through a longitudinal survey of graduate students and link it to career data, this would provide the most complete picture of how financial support is packaged and how it relates to graduate school and career outcomes. SRS should investigate whether it is possible and cost-effective to collect and link such data in that way.

SRS should also consider how it might augment data on the sources of graduate student support by linking SED data to data available on federal financial support of graduate students. This would include federally-funded fellowships and traineeships, and potentially, research assistantships. If principal investigators reported data on

graduate students supported by federal grants, SED data could be linked to these data, too.

Completion of Graduate School

Policymakers concerned about graduate education and the efficacy of certain types of funding have been seeking to know more about the factors affecting completion of graduate school. An internal NSF task force examining data availability for the analysis of graduate education argued "the proportion of graduate students who complete the curriculum they undertake is a very important intermediate outcome of federal support for graduate education" (NSF 1996). A recent NRC report on graduate school completion and attrition, *The Path to the Ph.D.*, also argued that while some attrition from graduate school is expected, proper interventions could encourage more students who do not complete to do so, and make contributions to our universities, science and medicine, industrial development, and society (NRC 1996a). Both the internal NSF report and the NRC report on attrition conclude that no national data exist that measure the completion rates of graduate students in all disciplines from the variety of institutions awarding doctoral degrees.

More complete data on the graduate school experience would allow better understanding of attrition and completion and of the interventions by government agencies or others that might best prepare Ph.D. students for their careers. Such data would include the decision to go to graduate school, career expectations, paths through graduate school to degree completion, financial support, reasons for attrition (if applicable), and the role of mentor, department, and career advising in completion/attrition and job placement. Data on these decisions, expectations, and relationships could potentially illuminate the graduate school experience for those grappling with policy issues in this area.

Completion and attrition data for racial and ethnic minorities would be particularly useful. A substantial amount of demographic data on gender, race/ethnicity, and disability status already exist and are used by SRS in a number of publications; other kinds of data—such as those on graduate completion or attrition—could provide insights on how to improve graduate school and career outcomes for minority groups. Data on minority attrition could inform policies and programs designed to retain students from minority groups in academic programs.

A survey that longitudinally tracks students through graduate school could potentially collect much of this data on completion and attrition at the national level though it would be expensive, as discussed above. It will also take additional time before data may be used in a longitudinal manner. Questions posed in such a survey could be fielded to students at the beginning of graduate studies, at formal entrance into a doctoral program, at the completion of all requirements but the dissertation, and at degree receipt.

Retrospective questions about career expectations and overall satisfaction with the respondent's Ph.D. program were added to a one-time module for recent Ph.D.s in the 1997 SDR. Analysis of these data may shed light on the kinds of additional data required in this area, and whether such data might be collected through a survey of graduate students or the SED.

A related issue in the progress of Ph.D. students through graduate school across disciplines is time to degree, which increased steadily over the last three decades before stabilizing in recent years (NRC 1996b). This issue is important both to graduate students, because of the opportunity costs associated with time spent in graduate school, and to graduate institutions, because of the costs required for supporting students over longer periods of time. Observers of graduate education have argued that lengthening time to

degree has a variety of causes including the need for longer training in the face of growing knowledge across fields. Some also argue that time to degree has been driven up by students remaining in graduate school to avoid a tough job market or by departments that benefit from the teaching and research duties carried out by advanced graduate students at low cost. Some have shown that the increase can be attributed to a "cohort effect" that occurs when entering classes are smaller than previous classes. This phenomenon logically results in more graduates coming from earlier cohorts.

More data on the progress of students through graduate school might better characterize this phenomenon. Again, such data could be obtained through a longitudinal survey of beginning graduates students if that were cost effective. Data on progress toward the doctoral degree could also be readily collected through the Survey of Earned Doctorates. For example, SRS could ask SED respondents for the date when they completed all requirements for the doctoral degree except for the dissertation. This could be used to disaggregate time to degree into pre-dissertation and dissertation phases.

Graduate Education and Career Skills

COSEPUP and the NSB have both argued that in a changing job market for Ph.D. scientists and engineers, where jobs and workplaces are more diverse than ever, graduate education needs to be reshaped to better prepare students for their future careers. COSEPUP noted that *"more than half of new graduates with Ph.D.s—and much more than half in some fields, such as chemistry and engineering—now find work in nonacademic settings."* It argued, therefore, that the narrow research focus of some graduate programs provided intellectual depth, but that it needed to be balanced by other skills required in the workplace (NAS 1995). COSEPUP recommended that the graduate education enterprise, particularly departments, implement reforms in the education of students in science and engineering who will work in either academic or nonacademic sectors. The NSB subsequently also argued that federal agencies should reward institutions that provide a range of educational options to Ph.D. students, including opportunities for interdisciplinary research and for learning teamwork, business management skills, and information technologies (NSB 1997).

SRS took steps to provide data on skills obtained in graduate school by adding a question to a one-time module on the 1997 Survey of Doctorate Recipients about the adequacy of training that respondents were provided by their doctoral programs in a range of knowledge and skill areas. Included were subject matter, general problem solving, oral communication, writing, quantitative skills, computers, teaching, teamwork, research integrity/ethics, networking, and management. We recommend that SRS utilize the data it has collected through the 1997 SDR to examine this issue to the extent possible.

We do not, however, recommend that SRS commit additional resources to collecting data in this area. The Ph.D. is and should remain a research degree. In our experience, graduate students are best served by developing subject matter expertise and having a meaningful research experience. There are also negative consequences for adding activities to promote "skills," such as lengthening time to degree. Also, many graduate students already improve communication skills through teaching assistantships and learn teamwork through research assistantships. We are skeptical that meaningful information can be collected on skills obtained in graduate school, or that data collected on skills obtained are likely to suggest how graduate programs might be reformed.

While we do not recommend that SRS commit additional resources to collecting data on "skills," SRS might consider obtaining data

on other factors or indicators related to graduate school and career outcomes. The possible collection of data on the decision to go to graduate school, career expectations, paths to degree completion, financial support, reasons for attrition (if applicable), and the role of mentor, department, and career advising in completion/attrition, and job placement have already been mentioned. Researchers would also like to be able to link student scores on the GRE to data from the Survey of Earned Doctorates and the Survey of Doctorate Recipients to examine further the predictive power of the GRE with regard to career outcomes.

Issues in the Science and Engineering Labor Market

Even before the difficulties in the labor market in the 1990s raised questions about the transition to employment of new and recent Ph.D.s, the 1989 NRC report *Surveying the Nation's Scientists and Engineers* argued that NSF needed to improve the data it provides on career paths and work of scientists and engineers. For example, the report recommended the collection of additional data describing key career transitions of scientists and engineers, such as entry into the labor force and mobility across fields and sectors. The report also recommended that NSF pursue the development of estimates for immigration to and emigration from the United States of scientists and engineers and that NSF include these estimates in its personnel data system (NRC 1989).

Despite the substantial technical work SRS conducted in developing a new personnel data system for the 1990s, calls for additional data on careers have continued. For example, in *Reshaping the Graduate Education of Scientists and Engineers*, the National Academies' Committee on Science, Engineering, and Public Policy (COSEPUP) concluded that "more information is needed on the career tracks followed by scientists and engineers both inside and outside universities" (NAS 1995). In its statement on *The Federal Role in Science and Engineering Graduate and Postdoctoral Education* the National Science Board recommended an exploration of "improved policy data ... to identify current and emerging national needs for the science and engineering workforce" (NSB 1997). To meet ongoing and evolving needs of policymakers and others for current information on the science and engineering labor market, SRS should improve the data it collects on the transition to employment, career paths of Ph.D.s, and the international flows of scientists and engineers.

Obtaining Improved Data on Science and Engineering Careers

The most comprehensive source of data on the science and engineering workforce is SRS's human resources surveys. Its three personnel surveys—the National Survey of College Graduates, the National Survey of Recent College Graduates, and the Survey of Doctorate Recipients—plus the integrated Scientists and Engineers Statistical Data System (SESTAT) that draws on them provides substantial data on scientists and engineers educated at the bachelor's degree level and higher in the United States.

The personnel surveys in the SESTAT system include data on the following individuals:

- Those with bachelor's or higher degrees in science and engineering who lived in the United States at the last decennial census (1990)
- Those with bachelor's or higher degrees in a non-science and engineering field who lived in the United States and worked in a science or engineering occupation in 1990

- Those with bachelor's or higher degrees in science and engineering from U.S. institutions since 1990

They do not include those who graduated with bachelor's or higher degrees in a non-science and engineering field since 1990 who now work in science and engineering. They also do not include those who received bachelor's or higher degrees in science and engineering who live in the United States, but received their degrees since 1990 from a non-U.S. institution, unless they are already included under one of the rules listed above.

The personnel surveys and the SESTAT system provides data that can be used to describe:

- Employment status and unemployment rates
- Characteristics of principal job, such as employment sector, occupation, work activities, salary, employment changes over the last two years, government support status, and relationship of principal job to degree
- Other career characteristics, such as membership in professional societies, work-related training, second jobs, (for Ph.D.s) postdoctoral positions held, and (for academically employed Ph.D.s) academic rank and tenure status
- Educational background, such as high school diploma, associate degree(s), first bachelor's, two most recent degrees, and degree fields
- Demographic characteristics, including age, gender, race/ethnicity, citizenship status, country of birth, disability, marital status, spouse's employment status, dependents, and parental education

The purpose of the SESTAT system, as seen in these data elements, has been to estimate the population of scientists and engineers in the United States and to characterize their employment and demographic patterns (NSF 1999k).

66

The personnel surveys also collect information through one-time modules on important current issues in the science and engineering labor market. For example, all three surveys asked in 1993 for data on respondents' labor force status in 1988. The 1995 SDR asked questions on work history and postdoctoral experiences of Ph.D. scientists and engineers. The 1995 NSCG and SDR added questions about professional output (articles, papers, and patents). All three surveys asked questions on alternative or temporary work experiences (i.e., consulting, contracting), the reasons for such work arrangements, and whether benefits were provided (NSF 1999k). The 1997 SDR also included retrospective questions on career expectations, satisfaction with the respondent's doctoral program, and characteristics of the job search posed to recent Ph.D.s (those earning a doctorate between 1990 and 1995).

There are other vehicles for collecting labor market information that SRS could tap further. For example, the Survey of Earned Doctorates, completed by new Ph.D.s at the time the doctorate is earned, poses a series of questions on postgraduation plans. These job market questions provide important trend data on type of postgraduation position (i.e, employment, further training), employment sector, and anticipated work activities. Other questions could be added to this section of the SED to make the data more robust. For example, as evidenced by the percentage of Ph.D.s who already have definite postgraduation plans at the time of degree receipt (62 percent), most Ph.D. scientists and engineers begin their search for employment or further training prior to graduation (NSF 1999i). Thus, questions could be posed to them about the job market search they engaged in prior to degree receipt. As another example, obtaining data on starting salary for those who do have definite commitments for employment would provide another indicator of the current status of the job market by field for new Ph.D.s and how this compares to the

job market for more experienced Ph.D. scientists and engineers. As will be discussed below, SRS may also draw on other sources for data on scientists and engineers such as professional societies and universities.

Creating and Refining the Science and Engineers Statistical Data System (SESTAT)

The SRS personnel surveys are the primary sources of data in the United States on the pool of scientists and engineers, and SRS has already implemented a series of changes to these surveys in the 1990s. In 1986 SRS asked the NRC's Committee on National Statistics (CNSTAT) to make recommendations and provide design specifications for a science and engineering personnel data system in the 1990s. CNSTAT assembled a study panel that issued a report in 1989 entitled *Surveying the Nation's Scientists and Engineers: A Data System for the 1990s.* Appendix C provides an overview of recommendations from this report that urged SRS to restructure, expand, and better integrate its three personnel surveys—the National Survey of College Graduates, the National Survey of Recent College Graduates, and the Survey of Doctorate Recipients—to meet the information needs of policymakers, planners, and researchers on the population of scientists and engineers in the United States.

SRS has won praise for the technical work it has performed on these surveys and for the development of the Scientists and Engineers Statistical Data System (SESTAT) following the guidelines of the 1989 NRC report. Individuals interviewed for this study believe that the design of the SESTAT system has effectively integrated survey results across the three surveys. Effective data collection by the Census Bureau, Westat, and formerly the NRC in administering these surveys during the 1990s, moreover, has led to low survey and item non-response rates for each of the surveys in the past. Questionnaire redesign has led to consistent responses across items and an expanded range of data collected on scientists and engineers and their careers.

The praise is well deserved, yet issues remain that need to be addressed in the labor market for scientists and engineers and the design of the SRS personnel surveys and the SESTAT system. We have not been charged with carrying out the "thorough, zero-based evaluation of the design and operation of [the SRS] personnel data system" at the end of the 1990s that the 1989 report recommended SRS conduct. Thus our observations do not provide an exhaustive review of the personnel data system and its design. SRS should carry out such a review in conjunction with the Special Emphasis Panel (i.e., advisory committee) of the Doctorate Data Project and other experts who may provide insight on the content and design of all three personnel surveys. However, we would like to provide general observations on the personnel surveys and other SRS human resources surveys that re-emphasize, modify, or augment the recommendations of the 1989 report in light of circumstances a decade later.

We would like to re-emphasize the 1989 recommendation that "NSF should increase the research utility of the science and engineering personnel data base by enriching the content of its surveys" by exploring three content areas that required monitoring then and still do today. First, SRS needs to modify or add content to provide greater understanding of "the career paths that scientists follow and the factors that influence key transitions, including initial entry into the labor force, mobility across fields and sectors, and retirement." Likewise, SRS should revise its personnel surveys to improve data on "the kinds of work that scientists do and how their work is changing in response to changes in technology, organizational structure, and other factors." Further, we recommend that SRS increase the research utility of the personnel data system by developing better estimates of the international flows of scientists and

engineers—the estimates of immigration and emigration that the 1989 panel urged SRS to pursue.

Transition to Employment for New Ph.D.s

In the early 1990s, new Ph.D.s in some fields encountered increasing problems in the job market. For example, the American Institute of Physics found that the percentage of new physics Ph.D.s still unemployed during the winter following degree receipt rose to 6 percent for Ph.D.s who received their degrees during the 1993-1994 academic year. It declined to 4 percent for those receiving their Ph.D.s in the 1994-1995 and 1995-1996 academic years, yet along with the decline in unemployment has come an increase in the percentage of physics Ph.D.s working outside the field of physics (Mulvey 1998). Similarly, the American Mathematical Association found that new Ph.D.s who were still unemployed in the fall following degree receipt reached an all-time high of 14.7 percent in 1995. This unemployment figure has declined since to 7.2 percent for 1997-1998 Ph.D.s, but this still remains higher than it was in the late 1980s when, for example, it was 5.7 percent for 1989-90 (Davis, Maxwell, and Remick 1998 and 1999).

Uncertain at the time whether this job market situation was a short-term problem or a sign of long-term structural change, the Association of Graduate Schools (AGS) associated with the Association of American Universities (AAU) identified the immediate sources of these job market difficulties:

It is a fact that the 1992-93 recession, the downsizing of industrial basic research labs, the tapering off of federal R&D funding as a consequence of the end of the Cold War, the influx of experienced scientists from the former Soviet Union, and the substantial budget reductions in many colleges and universities all had a deleterious effect on the employment of Ph.D.s, and they all hit at once (AAU 1995).

The future job market, the AGS argued, depended on the performance of the economy, future trends in federal R&D support, whether or not colleges and universities hired more faculty in light of a growing college age population, and whether the cut in industrial basic research was, itself, a short-term adjustment or a long-term structural change.

In light of job market problems such as these, COSEPUP undertook its study of graduate education. COSEPUP, the NSB, AAU, and others have raised the following kinds of questions about the transition of recent Ph.D.s to science and engineering employment and have called for additional data to address these questions:

- What are the labor market needs and opportunities for Ph.D.s, by sector, industry, and field?
- What are the job market expectations of new graduates?
- What positions are Ph.D.s obtaining in their early careers and what do recent patterns of postdoctoral positions and non-tenure track positions mean for careers of Ph.D.s?
- What is the unemployment rate for new Ph.D.s during the first year following graduation? During the first five years?
- What are appropriate measures of underemployment for this segment of the science and engineering doctorate population?
- What proportion of new Ph.D.s are working involuntarily out of field and how are increases in this proportion to be interpreted?
- What are the starting salaries of graduates, by type of post-graduation position?
- What are the characteristics of the search for a first job and how can the search be improved?

As AGS urged in 1995, "before we conclude that there is a long-term job crisis, we need to undertake the kind of information gathering that COSEPUP suggests and monitor development in the coming years" (AAU 1995).

While SRS is attuned to these issues for Ph.D.s generally, obtaining data on the job market for *recent* Ph.D.s is notoriously difficult. A serious gap in SRS data on science and engineering Ph.D.s has been the job market experience of Ph.D.s in the twelve months before and after receipt of the degree. While the SED captures postgraduate plans and status at the time the doctorate is awarded, it does not adequately capture information on the job search of Ph.D.s in the months before and after receipt of the doctorate. Meanwhile, the SDR does not pick up new Ph.D.s until at least nine months after they receive their Ph.D. Thus, it misses the period when Ph.D.s face the most uncertainty and greatest difficulty if they have not already obtained a definite commitment for work or further study at the time they receive their degrees.

SRS took steps to address this gap by fielding twenty survey questions on the job search in a one-time module in the 1997 SDR to be answered by those who received their doctorate degrees between July 1990 and June 1996. This cohort was asked to answer questions about:

- Career expectations (kind of work, employment sector) Ph.D.s had at the beginning of their doctoral programs
- The state of the job market at the time they completed their doctorates
- The time that elapsed between receipt of the Ph.D. and the time the doctorate recipient took a first career path job
- Constraints they encountered and resources they used in the job market
- Whether they found a job that met their career expectations and the relationship between the field of their degree and field of work

- The effect of completing the doctoral degree on salary, level of job responsibility, job security, degree of interest in position, degree of technical demand in work, management activities expected of them, and other aspects as specified by the respondent
- Overall satisfaction with the doctoral programs they completed

SRS is commended for taking this step to obtain data that address a critical policy issue. Though the data are retrospective and subject to some post hoc bias, the graduate education community eagerly awaits analysis of the data from the 1997 SDR on these issues.

The gap in data on the transition to employment, however, needs to be more fully addressed by SRS. While the new questions on the SDR noted above will hopefully yield valuable and interesting information, many of the questions are retrospective and past experiences may be described by respondents through the lens of the present rather than as they were experienced at the time. Because of this, we recommend that SRS take additional steps. SRS should add questions to the SED about the experiences of new Ph.D.s in the job market that they have had by the time of degree receipt. We believe that even with the addition of questions to the SDR, the SED should obtain at least a limited set of data on the job market expectations of new doctorate recipients as a counterbalance to the retrospective data that will be obtained at a later time. Also, adding a question to the SED on salary for those Ph.D.s who have a definite commitment would add important information about the status of employed Ph.D.s.

To supplement its own data SRS should continue its productive work with others to obtain data on the job market experience of new Ph.D.s in the immediate months following graduation. SRS has interacted extensively and productively with professional societies and the Commission on Professionals in Science and Technology (CPST) in

obtaining data on the job market experiences of new Ph.Ds. This interaction should be continued, and strengthened. SRS should also explore how it might work closely with colleges and universities to assist them with the development of standardized data sets on the placement of their recent graduates. In its recent report on graduate education, the AAU urged colleges and universities to maintain comprehensive data on completion rates, time-to-degree, and job placement for each of their graduate programs. The report specifically recommends that institutions track their graduates at least until first professional employment beyond postdoctoral appointments. The AAU suggests that institutions should provide student applicants with this information (AAU 1998). As research universities implement this recommendation, SRS should help them standardize their data collection efforts so that locally collected data could be aggregated in a meaningful way at the national level.

Career Paths of Scientists and Engineers

The NRC report *Trends in the Early Careers of Life Scientists* released last year has added further to the enumeration of problems facing recent Ph.D.s as they negotiate a changing labor market. The following passage from the report provides an overview of the problems facing recent Ph.D.s in the life sciences in particular:

The training and career prospects of a graduate student or postdoctoral fellow in the life sciences in 1998 are very different from what they were in the 1960s and 1970s. Today's life scientists will start graduate school when slightly older and take more than 2 years longer to obtain the Ph.D. degree. Today's life-science Ph.D. recipient will be an average of 32 years old. Furthermore, the new Ph.D. today is twice as likely as in earlier years to take a postdoctoral fellowship and thus join an ever-growing pool of postdoctoral fellows—now estimated to number about 20,000—who engage in research while obtaining further training and waiting to obtain permanent positions. It is not unusual for a trainee to spend 5 years— some more than 5 years—as a postdoctoral fellow. As a consequence of that long preparation, the average life scientist is likely to be 35-40 years old before obtaining his or her first permanent job (NRC 1998c).

Adding to the issues associated with a lengthier training period, the report identified problems with the job market:

The 42% increase in Ph.D. production between 1987 and 1996 [in the life sciences] was not accompanied by a parallel increase in employment opportunities, and recent graduates have increasingly found themselves in a "holding pattern" reflected in the increase in the fraction of young life scientists who after extensive postdoctoral apprenticeships still have not obtained permanent full-time positions in the life sciences (NRC 1998c).

This particular report drew heavily on SRS data to analyze the job market situation of life sciences Ph.D.s, and thus, provides an example of how useful SRS data may be. There still are questions about career patterns like this one that SRS data do not yet fully address. In his dissent to the report, for example, Henry W. Riecken cited the "totally inadequate evidential basis" for the report's recommendation that more federal resources should be funneled into training grants, because they would presumably enhance the career outcomes of the graduate students who receive them (Riecken in NRC 1998c). The job market situation for recent life sciences Ph.D.s, therefore, is an example of the kind of career issue that Ph.D.s, employers, and policymakers confront and for which they require additional data.

The range of appropriate data needed to carry out a valid comparative evaluation of the alternatives available to the federal government for supporting graduate students and the career outcomes of students with different types of support is incomplete. The need for improved data on financial support for students throughout their time in graduate school, one component of this analysis, was discussed earlier in this chapter.

To provide its users with data that allow broader and deeper analysis of career paths SRS should take a number of steps. First, SRS should strive to fully support longitudinal and time series analyses with its personnel data— especially data from the SDR. The current difficulties with using SDR data in these ways stem from several sources. In response to the 1989 NRC report, *Surveying the Nation's Scientists and Engineers*, SRS revised the SDR questionnaire in 1993 to make it more comparable with the other two personnel surveys and to expand the range of questions posed to respondents (Cox, Mitchell, and Moonesinghe 1998b). We endorse this questionnaire revision, but caution that while additional revisions have continued to be required to meet the needs of data users, SRS should avoid ongoing minor changes to questions that potentially disrupt time series. Second, longitudinal and time series analyses of the SDR have been compromised over the last decade because of "maintenance cuts" (i.e., changes in sample size due to fluctuations in survey budget) and new methods of survey follow-up (Cox, Mitchell, and Moonesinghe 1998a and 1999b). There is no doubt that improved survey follow-up techniques (e.g., computer-assisted telephone interviewing) have improved response in such areas as race, have compensated for earlier non-response bias on such variables, and have thus contributed to survey quality. SRS, however, should avoid changes in survey samples from year-to-year that compromise the longitudinality of surveys like the SDR, obtained at great expense and with a respondent burden that is hard to justify if the data cannot be used longitudinally.

Because SRS has adopted high statistical standards in the 1990s, it has decided to eschew longitudinal and time-series analyses of SDR data for the reasons stated above. While we applaud its focus on data quality and statistical standards with respect to data collected from this point forward, we urge SRS not to err on the side of caution with respect to the data already collected. Analysis of these unique longitudinal and time series data from the 1970s and 1980s must not be abandoned simply because they are less than perfect. SRS should support analysts in their use of these data.

Second, evidence from the interviews conducted for this study suggest that SRS could provide better career path data by making it available at a more detailed level as well. Often important trends in the labor market are field or sub-field specific. Thus, in order to properly analyze a labor market issue and reach conclusions as to what, if any, policy adjustments need to be made, analyses need to be carried out at this sub-field level. SRS should, along with its data users, consider options for addressing this problem. One option might be to increase the sample size for the SDR. Because the SDR samples only 8 percent of the doctorate-level scientist and engineer population, cell sizes are too small to be used in analysis. SRS should consider increasing the sample size to facilitate this level of analysis. We recognize that this could be an expensive fix for the problem; however, we suggest that SRS consider this and other means of addressing this analytic issue in a cost-effective manner.

Third, NSF should work with the National Endowment for the Humanities (NEH) and private foundations to revive the humanities component of the Survey of Doctorate Recipients. This component, administered through the Survey of Humanities Doctorates, provided data on the careers of Ph.D.s in

humanities fields (history, art history, philosophy, English and American language and literature, modern language and literature, classics, music, and other humanities fields). This survey was fielded biennially by the National Research Council with funding from NEH, through NSF, as a component of the Longitudinal Doctorates Project from 1977 to 1995.

The loss of humanities data from the SDR represents one of the biggest gaps in the data on the academic sector of which academic science and engineering is a part. The data are important to humanists concerned about career and labor market trends in their fields. The data are also important to analysts who seek to look comprehensively at research universities and the career paths of academically employed doctorate recipients. This is so for two reasons. First, to fully understand the health of academic science and engineering it is important to understand the health of the entire academic enterprise. Second, analysis of trends in academic careers of scientists and engineers is richer when trends in careers of academic humanists are also available for analysis. From the perspective of those interested in scientists and engineers, humanists provide a comparison group that allows analysts to determine whether certain academic career trends are specific to science and engineering fields or occur across all fields. For example, the growing number of postdoctoral fellows is seen in science fields, particularly the life sciences, and far less in the humanities. Also, when those who work part-time are asked why they do so, 43 percent of humanists working part-time in 1995 responded that there was "no suitable job available" compared to just 22 percent of scientists and engineers working part-time (NRC 1997b; NRC 1998a). These kinds of differences provide important comparative information that illuminates trends across science, engineering, and the humanities.

Again, we recognize the substantial costs associated with reviving the humanities SDR.

We also recognize that the NEH budget was substantially reduced in 1995 and has remained flat since that time. However, NSF could approach private foundations as well as NEH for funding. Also, while we prefer this component to be restored on a biennial basis, we recognize that having humanities data on a quadrennial basis beginning with the 2001 SDR is better than having none at all, and would support this less-costly option for reviving these data.

Fourth, NSF should investigate the cost-effectiveness of linking its personnel data (especially the SDR) to data on grants, publications, and patents to facilitate deeper investigation of career outcomes. For example, federal agencies that provide research grants keep electronic records of their grant recipients, and research award data could be linked to SDR records. In 1997, for example, the SDR oversampled individuals who received NSF Graduate Research Fellowships (NSFGRF). Data on these individuals could be linked to data from the NSFGRF in an ongoing evaluation of that program, but we would also like to see SDR data linked to these kinds of data on federal awards supporting education or research for use as part of the SDR data file. For some agencies, these award data have been incomplete, e.g., only the principal investigator (PI) is listed for research grants and not all associated faculty. Still, such linking could enrich the SDR as a data set. Linking personnel data with publication and patent data is more difficult, but SRS should seek ways to make such linkages easier. SRS asked for data on the number of articles, publications, patents applications, and patent awards in a special module for the 1995 NSCG and 1995 SDR. If these data points were collected on a continuous basis they would provide benchmarks against which researchers could measure their ability to link SRS with publication and patent databases.

Nonacademic Careers

To better understand the career paths of scientists and engineers and the career options of new Ph.D.s, SRS must place a high priority on revising the SDR to obtain data that better describe nonacademic careers of Ph.D. scientists and engineers. The SDR questionnaire is oriented too much toward surveying doctoral scientists and engineers who hold faculty positions in colleges or universities and not enough toward those who work in government, business, or nonprofits.

The majority of doctoral scientists and engineers do not work for educational institutions, but for private businesses, government agencies, or nonprofit organizations. In 1995, 48.5 percent of doctoral scientists and engineers worked in educational institutions. Of these, 42.6 percent worked in 2- or 4-year colleges or universities, 4.7 percent worked in university-affiliated research institutions, and 1.1 percent held teaching positions in elementary or secondary schools. Of those who were faculty and teachers, moreover, only about three-quarters were in tenured positions or tenure-track jobs. Just one-third of all employed, doctoral-level scientists and engineers fit the stereotype of tenure-track or tenured faculty (NRC 1998a).

This has not always been the case. While the percentage of Ph.D. scientists and engineers working in educational institutions has been declining since the SDR was first fielded in 1973, a majority nonetheless did so until the 1990s (NSF 1991). As shown in Table 4-2, this labor market indicator first fell below 50 percent in the 1990s. Since it did so in the midst of job market difficulties for new Ph.D.s, this trend was finally given the level of attention it had deserved for some time. The COSEPUP report, for example, made much of

Table 4-2 Employed Doctoral Scientists and Engineers by Sector of Employment, 1987-1995

Sector	1987	1989	1991	1993	1995
Educational institutions	51.7	51.3	46.9	47.9	48.5
Industry	32.0	32.5	36.2	36.6	36.2
Government	9.2	9.0	9.1	10.1	9.9
Nonprofit	6.6	6.6	6.8	5.1	4.9
Other	0.4	0.4	0.4	0.3	0.6
No report	0.2	0.2	0.5	*	*

*Beginning with 1993 missing data are imputed.

Source: Cox, Mitchell, and Moonesinghe 1998b, p. IV-8.

the fact that the majority of Ph.D. scientists and engineers work outside of academia, and thus called for more information on the career paths of these Ph.D.s both inside and outside of colleges and universities.

SRS could better describe the kinds of careers pursued by Ph.D.s outside of academia by adding to the SDR questions that solicit the following kinds of detail to flesh out these careers:

- Data on compensation of scientists and engineers, particularly in the private sector. The SDR currently asks for salary, but this does not adequately describe compensation in the private sector where it may include stock options, bonuses, and additional benefits.

- Data on the productivity of Ph.D.s in the private sector where they may focus as much on innovation as knowledge generation. In its 1995 cycle, the SDR asked for information about articles, papers, and patents, and these are valuable data points, but these may provide an incomplete measure of scientific productivity for a researcher in a private firm that guards its scientific advances and innovations as proprietary information. In this instance, an expanded description of compensation may be the best proxy for

productivity in the private sector although SRS should consider the range of options for capturing productivity in the private sector.

- Data on new occupations and not easily identifiable research jobs for Ph.D.s in industry (especially in emerging technology-based fields such as biotechnology and information technology) in such areas as sales, regulation, and patenting. This was a subject of much discussion at the September 1998 workshop. There was little consensus on whether it is necessary to train people to the Ph.D. level for these positions. However, participants noted that the number of individuals in such positions is growing and that this phenomenon requires tracking and better understanding of tasks performed.

This is not an exhaustive, but rather an illustrative list of the kinds of data that could be collected. For each of these areas, new data would shed light on developments that weigh on the thinking of policymakers and program administrators who seek to assure that new degree recipients may have successful outcomes in industrial and other nonacademic careers, as well as academic careers.

Work Arrangements, Field, and Occupation

Surveying the Nation's Scientists and Engineers recommended that "NSF should assign priority to new or modified content items that will provide greater understanding of the kinds of work that scientists and engineers do and how their work is changing" (NRC 1989). This recommendation holds as true today as it did a decade ago, not because SRS did not address this in the redesign of its personnel survey instruments in 1993, but because the science and engineering enterprise continues to change. Specific kinds of data about the work of scientists and engineers that analysts would like to have today include:

- The kinds of positions they hold (e.g., permanent/temporary, full-time/part-time, contract/consultancy, job sharing; etc.)

- The *nature* of other jobs or positions held simultaneously, particularly when the additional position is in a different organization as in the case of a university faculty member also working in a start-up company (e.g., administrative positions, supplementary employment, consultancies, start-up companies, joint appointments at other institutions)

- The organization of the work they engage in (e.g., traditional organization, ad hoc teams, virtual teams, inter-organization partnerships, consulting/outsourcing)

- Job and career flexibility (e.g., flex-time, telecommuting, portable benefits)

SRS has already begun to collect data on these kinds of new or alternative work arrangements. In the 1997 survey cycle, the personnel surveys include a special module on alternative or temporary work arrangements, such as contracting or consultancies. SRS, in concert with its data users, should continue to examine how all three of its personnel surveys answer these questions today and revise or add content items as needed to better describe the

current work arrangements of scientists and engineers educated at the bachelor's degree level and above.

Another issue that should be better addressed in the SRS personnel surveys concerns whether appropriate data are being collected on field and occupation. *Surveying the Nation's Scientists and Engineers* recommended that "key questionnaire items in the SDR be made comparable with those in the NSF Panel Survey. Specifically, it is imperative to include a question on occupation in both surveys that does not bias respondents toward reporting their degree fields and that conforms with the [Standard Occupation Classification]" (NRC 1989). In response, SRS dropped field of science and engineering from the SDR and added questions to obtain occupation codes utilized by the Bureau of Labor Statistics that were designed to better capture the kinds of positions held by scientists and engineers. This change standardized SRS occupational data with the other personnel surveys.

To further refine the data collected in this area, SRS should consider adding current field of science and engineering back into the SDR questionnaire while also retaining the question added on occupation. The occupational categories added to the personnel surveys allow for better integration of these data with other workforce surveys. However, the work of scientists and engineers cannot be adequately described and their careers tracked without data on the field in which they currently work. Analysts need field data to more fully characterize the work of scientists and engineers. For example, respondents who may categorize themselves as "managers" may yet define themselves as working in a specific field, such as "chemistry." Analysts also need to know if a respondent has switched fields or is working in more than one field. Given the inter- and multi-disciplinary nature of research, it would also be advisable to allow respondents to the SDR in particular to select more than one field in which they are working.

International Flows of Scientists and Engineers

Surveying the Nation's Scientists and Engineers also argued that "NSF should pursue its planned research program to develop estimates of immigration and emigration of scientists and engineers and to develop ways of incorporating such estimates into the personnel data system" (NRC 1989). The immigration of scientists and engineers has long been a feature of the U.S. science and engineering enterprise. Data on immigration and emigration suggest, however, that flows of scientists and engineers into and out of the United States have only intensified in the last ten years, so the recommendation to develop and improve estimates of immigration and emigration is, if anything, of greater import today than in 1989. Some indicators of increased international flow in the science and engineering labor market include the following:

- In 1993, foreign-born individuals accounted for 16 percent of all scientists and engineers and 29 percent of doctorate-level scientists and engineers engaged in R&D in the United States (NSF 1998c).

- From 1986 to 1996, non-U.S. citizens grew from 22 to 33 percent of individuals receiving doctorates from U.S. colleges and universities. The number and percentage of non-U.S. citizens, however, varied by field and is higher in the natural sciences and engineering: in 1996, non-U.S. citizens made up 58 percent of new engineering Ph.D.s, and 47 percent of new Ph.D.s in each of the life sciences and physical sciences (NRC 1998b).

- Non-U.S. citizens accounted for two-thirds of the growth in the number of Ph.D.s from 1986 to 1996. The number of non-U.S. citizens receiving life sciences Ph.D.s increased 184 percent during that ten-year period (NRC 1998b).

- Citizens of China, India, Taiwan, and Korea made up 55 percent of the non-U.S. citizens who received Ph.D.s in the U.S. in 1995. Citizens of China have increased the rate at which they stay in the United States after graduation so that, in 1995, 96 percent planned to remain in this country. Similarly about 90 percent of Indian citizens have planned to stay in the United States. By contrast, increasing percentages of Ph.D.s from Taiwan and Korea plan to return to their country of origin after degree receipt (NRC 1996b).

The science and engineering enterprise has become global in its dimensions in the 1990s and so has the science and engineering labor market.

While the flow of funding for research and development is as important to the science and engineering enterprise as the flow of people, comparability of data across nations is weaker in the human resources area than for R&D funding. Since the movement of personnel is one of the principal ways to both facilitate the diffusion of knowledge and to meet demand for skills in the labor market, obtaining better data in this area should be a high priority for SRS and the first step would be to develop a strategy for improving its collection and acquisition over time. Substantial work on how to produce and acquire data both on comparisons of national populations of scientists and engineers and on international "flows" or "circulation" of scientists and engineers is needed. We recommend that NSF provide SRS the resources to continue and expand significantly its data collection and analysis in this area.

Analytical issues on the international flow of scientists and engineers that could be informed by the additional data include:

- The number of U.S. citizens who study abroad and why they do so

- The number of foreign-educated scientists and engineers in the United States, by citizenship status, field and occupation, and whether they plan to stay

- The contribution that non-U.S. citizens who earned Ph.D.s in the United States have made to their home countries in teaching, research, and technology development

- The patterns of circulation that are of benefit to the United States, home countries, and to the global diffusion of scientific knowledge

It is important to an analysis of science and engineering resources, their distribution, and use, that these issues be understood. While obtaining these data is not an easy matter, they can help illuminate science resources issues.

SRS should develop a strategy for improving its data on the international flow of scientists and engineers and we urge NSF to fund additional data collection in this area. SRS should begin by focusing particular attention on improving estimates of immigrant and non-immigrant foreign-trained scientists and engineers, foreign students, and postdoctoral fellows in the United States.

Data Users and External Researchers

The Panel that wrote *Surveying the Nation's Scientists and Engineers* argued that the personnel data system in the 1990s should "provide a research base for improved analysis of relevant labor markets and of flows into, out of, and within the science and engineering labor force that can pinpoint trouble spots and provide early warnings of future problems, and ...support basic innovative research on scientists and engineers and the science and engineering pipeline." SRS has performed well in creating a science and engineering

personnel data system for the 1990s. Today, the SESTAT system and the three personnel surveys from which it draws data provide solid information for describing the stock of scientists and engineers who were educated and work in the United States. To create an improved personnel system for 2000 and beyond, SRS will also have to provide data on the flows of individuals through career paths and transitions, across fields and sectors, and across national boundaries.

To accomplish this, SRS needs to develop a comprehensive plan for the SESTAT data system for the decade beginning in 2000 that takes these flows into account. The plan needs to specify analysis goals that can be used to guide both survey and sample designs. SRS should begin its work on "SESTAT 2000" with a research statement for each of the three personnel surveys that contribute to SESTAT. These research statements should detail specific policy questions the data collected from each survey are designed to address. These statements may be amended as new issues warrant. In the meantime, they will strengthen the surveys by focusing them on important issues.

To accommodate new data needs SRS should re-examine survey content for the three personnel surveys and make tradeoffs on specific questions to be asked. The questionnaires for the surveys were expanded beginning in 1993 to capture additional information, and are now already long and expensive. Additional questions need to be added at this time, but increasing survey length further will increase respondent burden and cost. Some current questions may be retired to make room on the questionnaire for new questions, though SRS may also field questions on some new issues through modules that change from cycle to cycle, thus allowing additional questions to be asked on a one-time basis.

In designing SESTAT 2000, SRS should also consider the frame from which it draws its samples. Currently, the three personnel surveys do not include, and therefore do not provide data on, scientists and engineers in the United States who did not receive degrees from U.S. institutions except those in the United States at the time of the 1990 census. They therefore omit a potentially substantial segment of the science and engineering workforce. A discussion of additional alternative sampling frames for the personnel surveys is included in Appendix D.

We would like, finally, to re-emphasize two more of the recommendations made by the 1989 panel. First, this panel argued that "NSF should actively solicit feedback from its users on the design, content, and quality of the data system, and on the content and format of data products. NSF should consider for this purpose establishing a user panel to provide input on a regular basis." This remains an important recommendation today. SRS has recently organized a Special Emphasis Panel for the Doctorate Data Project that includes the SED and SDR. We recommend that this panel also assist with the design and content of all three personnel surveys in SESTAT for 2000 and beyond.

Second, the 1989 panel also argued that "NSF should actively encourage and provide support to researchers for innovative studies of science and engineering personnel using survey microdata. NSF should consider for this purpose establishing a grants program to fund projects that use the personnel data." This sentiment was also echoed in the 1995 COSEPUP report, *Reshaping the Graduate Education of Scientists and Engineers*. Again, this recommendation still needs to be more fully implemented. Indeed, SRS data are often underutilized and nowhere is this truer than for the data obtained through the three personnel surveys that provide data to the SESTAT System. The SDR is a prime example. From January to October 1997, the National Research Council, then administrator of the surveys, received just seven requests for custom tabulations from the SDR compared to 27 such requests made for data from the SED.[1]

SRS should better publicize SESTAT data and also allow SESTAT data to be more available to and accessible by external researchers. As noted in Chapter 3, SRS and its data would benefit from a program that provides grants to external researchers to utilize SRS data in their analyses. Within this program external researchers who utilize SESTAT data should be given special consideration, so that this underutilized database receive additional use, scrutiny, and exposure. These researchers should also be involved in writing the research statements that will guide SESTAT as SRS revises this survey for the next decade and specifies the kinds of analyses that will be performed utilizing data from the personnel surveys.

Finally, SRS should monitor and summarize research conducted by others using data from the NSCG, NSRCG, and the SDR. This could be tasked to a contractor responsible for research. Other federal statistical agencies provide similar summaries of research based on data of use to social scientists. If SRS were to do this also, such summaries would be an aid to researchers as well as a source of information for SRS in its role of advising policymakers.

---

[1] These are requests for custom tabulations and do not include requests that were handled with existing data or tabulations via phone call, e-mail, or fax.

5

Research and Development Statistics: Funding, Performance, and Innovation

The Changing Organization of R&D

Changes in the sources and organization of research and development (R&D) funding and performance also prompt a review of the SRS portfolio at this time. Indeed, many structural changes in R&D have been underway since the early 1980s, so a review of the R&D data program is long overdue. The need to address these changes is substantial if we are to fully understand the allocation and use of science and engineering resources and the ways they contribute to national goals.

R&D in the Post-World War II Era

The use of scientific and technological R&D during World War II initiated a new era in the history of the financial resources made available for scientific research and technological development in the United States. Before World War II, basic scientific research evolved in universities independently of the federal government, often with philanthropic support. Industry funded its own scientific and technical research, often in central laboratories. The scientific and technical requirements of the military in WWII changed this as the federal government engaged the science and engineering enterprise in the production of weapons systems in support of the war effort. Towards the end of WWII, Vannevar Bush, director of the wartime Office of Scientific Research and Development, saw that successful coordination of scientific research to solve the war effort's technical issues could be profitably continued in peacetime to promote the nation's health, economic well-being, and national security. While beginning his work under the Roosevelt administration, Bush delivered to President Harry Truman his 1945 report, *Science: The Endless Frontier*, which provided the rationale and targets for federal investments in a federally-supported science and engineering enterprise that has continued into the 1990s (Bush 1945).

While Bush's report urged the use of science to support health research and economic progress, the Cold War that soon followed WWII made defense a substantial component of the federal role in scientific and technological research. During the ensuing years, the defense rationale for the federal role only intensified. As is well known, the Soviet launch of Sputnik in 1957 shocked the United States into additional investments in U.S. science. This led to the creation of the National Aeronautics and Space Administration (NASA) and spurred the American drive to place a man on the moon by the end of the 1960s. In the aftermath of Sputnik, the United States also created the Defense Advanced Research Projects Agency (DARPA) (Boesman 1997).

These events set the stage for postwar funding of science and engineering in the United States. When data were first collected by NSF on sources of R&D funding in 1953, the federal government was already the leading source of funds for R&D in the United States and this continued to be the case for the next three decades (NSF 1999b). Similarly, federal spending on defense R&D exceeded federal non-defense R&D spending from the early 1950s through the mid-1960s when civilian R&D finally caught up due to explosive increases in space science. Defense and non-defense R&D were then funded at relatively comparable levels until the early 1980s (AAAS 1999).

U.S. R&D Since 1980

Total annual R&D expenditures in the United States by industry, government, universities and colleges, and nonprofit organizations reached a record high of $220.6 billion in 1998. Adjusted for inflation, R&D expenditures increased 5.3 percent from 1997 to 1998. Moreover, with this new level of spending, R&D as a percentage of gross domestic product (GDP) is 2.61 percent in 1998, the highest level since 1992 (NSF 1999f).

These expenditure levels are within the bounds of historic trends. Fueled by continued increases in industrial spending on R&D, average annual growth in total U.S. R&D expenditures in real terms has been 4.3 percent per year for the period 1995 to 1998. This is substantially higher than the 1.6 percent average real annual growth in total U.S. R&D for 1985 to 1995, but also lower than the 5.6 percent real average increases for the period 1975 to 1985. Also, R&D as a percentage of GDP increased recently, but it remains below R&D's share of GDP in the early 1990s when it exceeded 2.7 percent. The 1998 figure of 2.61 percent falls between the low point in the last 40 years, 2.12 percent in 1978, and the high point, 2.87 percent in 1964 (NSF 1999f).

While aggregate levels of the nation's R&D spending are within historic bounds, policymakers and researchers have observed that structural changes within the U.S. R&D system have been occurring for some time and will likely continue well into the future (NSB 1998a, Mowery 1999). These changes can be seen in industrial R&D and innovation, federal R&D funding, and academically-performed R&D.

Industrial R&D and Innovation

Changes in industrial R&D have been a central factor in the restructuring of the U.S. R&D system generally. There have been substantial increases in industry R&D expenditures overall, sizeable increases in the proportion of industrial R&D performed in the service sector, and changes in the structural organization of industrial R&D and innovation.

First, industry has been the largest source of R&D funding in the United States for almost two decades, and its share continues to grow. In 1980 industry passed the federal government as the largest supplier of R&D

funding in the United States. Industry's share has continued to grow such that in 1998, industry expenditures on R&D reached $143.7 billion or 65.1 percent of total R&D expenditures. The growth in industry's share of U.S. R&D is the result of continued, high-rate growth in industry R&D spending and decreasing R&D spending by the federal government. The average real annual growth in industry R&D spending from 1977 to 1997 was 5.4 percent. After a decline in real industry R&D spending earlier in the 1990s, average real annual growth in industry R&D spending from 1994 to 1998 was 6.2 percent. Meanwhile, federal R&D expenditures, which grew 60 percent in real terms from 1975 to 1987 and closely tracked the aggregate amount of industry spending during that time, has since declined by more than 20 percent in constant dollars.

An important result of the declining defense share in federal and national R&D funding has been a shift in funding for development from government to industry. As shown in Figure 5-1, development expenditures by industry and the federal government as a percent of total R&D expenditures from these sources has been stable over the past quarter century. However, the share of development funding from these sources provided by industry has increased from one-half to three-quarters. Industry spending on development has increased substantially, while federal spending on development, especially defense development, has declined (NSB 1998a).

Indicators from the mid-1990s suggest that trends in sectoral share of expenditures would be the opposite for basic research, but this has not been borne out in the long run. In 1991, industry provided one-third of the combined spending by industry and the federal government on basic research. Industry spending on basic research dropped, in real terms, from 1991 to 1992, and then held steady until dropping again in 1995, but

industry spending on basic research has since rebounded. In 1997, the combined level of funding by industry and the federal government for basic research is about the same, in constant dollars, as in 1991, and the relative contributions of the federal government and industry are also about the same now as in 1991. With some minor fluctuation, industry has allocated its expenditures across the categories of basic research, applied research, and development in a fairly consistent manner over the past quarter century. For example, in 1977 basic research, applied research and development comprised 4.4 percent, 22.5 percent, and 73.2 percent of industry spending respectively. In 1997, the respective shares for these categories were 6.0, 22.1, and 71.9 percent (NSB 1998a).

Second, the distribution of industry R&D performance across industrial sectors has changed substantially over time, prompting the National Science Board to write, "probably the most striking change in industrial R&D performance during the past decade is the service sector's increased prominence" (NSB 1998a). An important component of service sector R&D that has been driving this increase is software development. In the early 1980s, R&D performance by non-manufacturing industries made up less than 5 percent of total industry R&D performance. Then, R&D performance by non-manufacturing firms as measured by SRS doubled between 1984 and 1988 and doubled again between 1988 and 1990 as seen in Figure 5-2. By 1991, non-manufacturing industries accounted for almost 25 percent of industrial R&D performance. While some of this apparent growth is due to the reclassification of some R&D (e.g., in software and telecommunications) from manufacturing to non-manufacturing sectors, this expansion in the percentage of R&D in the service sector corresponds with a similar expansion of the service sector within the economy generally.

Third, the structure of industrial R&D and innovation has been reorganized in the last two decades. In the early 1980s the competitiveness and technological leadership of U.S. firms in a global economy were generally seen as troubled at best and imperiled at worst (NRC 1999; Mowery 1999). A report by the NRC's Board on Science, Technology, and Economic Policy (STEP), *Securing America's Industrial Strength*, has outlined the factors generating resurgence in U.S. industrial strength amidst increasing global competition and technological change from the early 1980s to the late 1990s. Elements of the resurgence have included federal policy changes, shifting firm strategies, and changes in the U.S. system of innovation.

Changes in federal macro- and microeconomic policies since 1978 have created a new economic and regulatory environment that has encouraged U.S. industrial resurgence. Macroeconomic

policies in recent years have helped keep inflation low and economic growth strong. Beginning with airline deregulation during the Carter administration in 1978, a series of changes in microeconomic policies have also contributed to U.S. industrial resurgence. These policies have included economic deregulation, lenient antitrust enforcement, protecting intellectual property, trade liberalization, and ongoing federal support of research and development (NRC 1999).

At the same time, firms have pursued strategies that have also been critical to industrial resurgence. A particularly important source of renewed strength for U.S. firms has been their ability to reposition themselves in their markets by introducing new products and processes. *Securing America's Industrial Strength* notes that firms have accomplished this by "capitalizing on shifts in demand to create new markets, often through the deployment of technologies new to those industries, as well as to accomplish

Figure 5-1 Distribution of Combined Industry and Federal Agency Spending on Research and Development across R&D Categories, 1975-1997

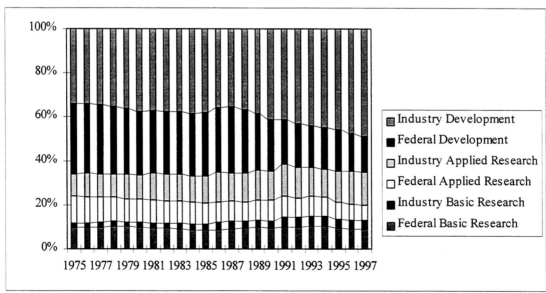

Source: National Science Board, *Science and Engineering Indicators—1998,* Appendix Tables 4-10, 4-14, 4-18 (1996 and 1997 data are preliminary).

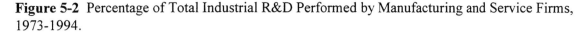

Figure 5-2 Percentage of Total Industrial R&D Performed by Manufacturing and Service Firms, 1973-1994.

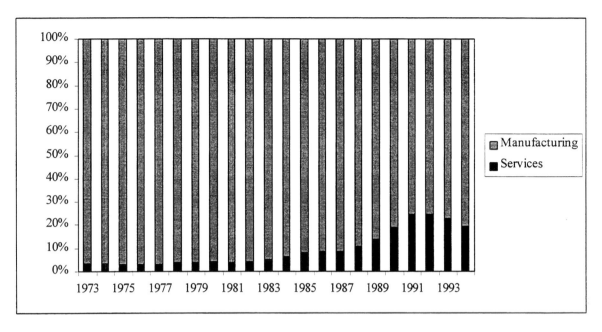

Source: National Science Board, *Science and Engineering Indicators—1998,* Appendix Table 6-9.

cost reduction and quality improvement" (NRC 1999). Other firm strategies have included merging with or acquiring other firms, focusing on a specific product niche, and globalization of R&D, production, and distribution.

Perhaps more to the point, a changing organization of innovation has also contributed to the improved position of U.S. firms in domestic and global markets. *Securing America's Industrial Strength* illustrates the point by citing Timothy Bresnahan's description of how innovation in the computer industry has been reorganized from an "IBM" to a "Silicon Valley" system of innovation. Under the IBM model, innovation was "integrated, hierarchical, [and] more self-contained." In Silicon Valley, by contrast, innovation in the computer industry is organized through co-location of "multiple innovative companies excelling in components, hardware, software, networking,

and other specialized parts of the industry" (NRC 1999).

Increased collaboration in R&D has played a growing role in the organization of the U.S. innovation system. A series of legislative acts has encouraged this collaboration. For example, the National Cooperative Research Act of 1984 (NCRA) has provided firms that collaborate on "generic, pre-competitive research" with certain protections in the event of private suits and limited damages in the event that cooperative R&D is later found to be in violation of anti-trust statutes. Research Joint Ventures (RJVs) registered under the NCRA with the Department of Justice have totaled more than 665 from 1985 to 1996 (NSB 1998a). In addition, many firms, beginning with some large corporations with central research facilities, have reduced internal R&D capabilities and have relied increasingly on outsourcing. This has been true of such large firms as AT&T, IBM, Dupont, Xerox, and General Electric during the mid-1990s, but not of other large corporations, such as the Ford Motor Company. Even so, it is not yet clear whether this is a long-term trend or a one-time structural adjustment. In general, there has been a "proliferation of joint research ventures, strategic alliances, with foreign and other U.S. firms, ...and cooperative arrangements with federal laboratories through CRADAs"[1] (NRC 1999).

Federal R&D Expenditures

Meanwhile, shifting patterns in federal spending on research and development since the 1960s have been shaped by changing federal policy priorities and consequent changes in funding provided to mission agencies.

As shown in Figure 5-3, federal expenditures on defense R&D and non-defense R&D were roughly comparable from the mid-1960s until the early 1980s. Then, a rapid build-up of defense R&D spending in the 1980s combined with cuts in non-defense R&D set the pattern for nearly the rest of the decade. Defense R&D expenditures were reduced slightly in 1990, and with the end of the Cold War, have been reduced further, while funding for non-defense R&D has rebounded. The President's fiscal year 2000 budget proposed to spend more on non-defense R&D than defense R&D for the first time since 1981.

The most discussed trend in civilian R&D this decade has been the accelerated spending on health-related R&D, especially through the National Institutes of Health (NIH). This is the latest, though potentially most enduring, of a series of trends that have resulted from changing civilian policy and budget priorities. As shown in Figure 5-4, funding for the U.S.

space program in the 1960s resulted in substantial space-related R&D during that decade. By the 1970s, space R&D funding had declined, but reaction to the energy crisis in the wake of the oil embargo by the Organization of Petroleum Exporting Countries (OPEC) led to substantial increases in energy-related R&D that continued into the 1980s. As defense R&D expanded in the 1980s, space R&D was cut further, and then energy R&D funding was reduced. Space science funding has rebounded somewhat in the 1990s, but the primary trend of the most recent decade has been the acceleration of the long-term growth in funding for health-related R&D. A less observed, but also important trend has been the increase in general science funding, including budget increases for the National Science Foundation that have been used across fields of science and engineering R&D.

These changes in federal spending priorities have had consequences for expenditures across R&D spending categories. The defense build-up of the 1980s generated a bulge in federal development expenditures. In constant dollars, federal development expenditures increased almost 50 percent from 1980 to 1988 before declining again to the 1980 level in 1996. Similarly, federal expenditures on applied research increased more than 40 percent from 1980 to 1990, but the subsequent decline has not been as steep as for development. Federal applied research expenditures remain above the 1980 level. Federal expenditures on basic research, however, increased steadily in constant dollars from 1980 to 1994, before taking a two-year decline. Federal expenditures for basic research have since increased in fiscal years 1997, 1998, and 1999.

---

[1] CRADAs are Cooperative Research and Development Agreements.

Figure 5-3 Federally-Funded R&D for National Defense and Civilian Functions, Fiscal Years 1955-2000 (millions of constant 1992 dollars).

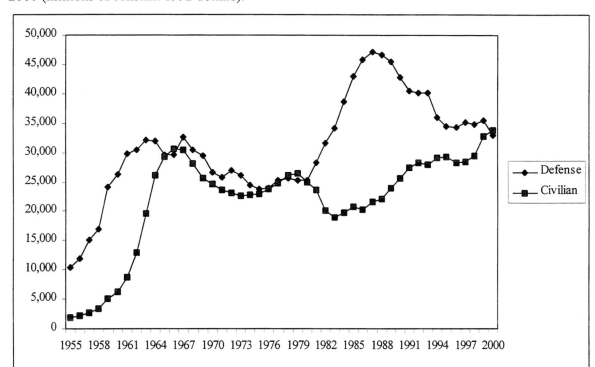

Note: Fiscal year 1999 is the latest estimate and Fiscal year 2000 is based on President's budget request. Figures for both fiscal years are estimated based on data in AAAS, converted to 1992 constant dollars using deflators in President's Budget Request, Fiscal Year 2000.

Sources: NSF 1999d, Table 3; AAAS 1999, Table I-16.

Changing priorities have also had an impact on funding for specific fields of science and engineering. A recent analysis by the NRC's Board on Science, Technology, and Economic Policy (STEP) revealed that, in the 1990s, increases in funding for the National Institutes of Health and other mission agencies along with cuts in defense spending may be creating long-term increases in some fields and decreases in others. For example, between 1993 and 1997, overall federal spending increased in the biological sciences, medical sciences, aeronautical engineering, computer sciences, materials engineering, and oceanography. Federal spending decreased in electrical engineering, mechanical engineering, physics, mathematics, chemistry, chemical engineering, civil engineering, and geology (NRC 1999).

Academically-Performed R&D

Analysts have also been concerned about the impact of these changes on the portfolio of R&D activities performed by the nation's colleges and universities. While the federal government provides the majority of funding for academically-performed R&D, its percentage of such R&D declined from 70 percent in 1970 to 58 percent in 1991. It has hovered since at around 59 or 60 percent. The share of state government funding for academically-performed R&D has also

Figure 5-4 Federal Civilian Research and Development Funds, by Budget Function, Fiscal Years 1961-1998 (millions of constant 1992 dollars)

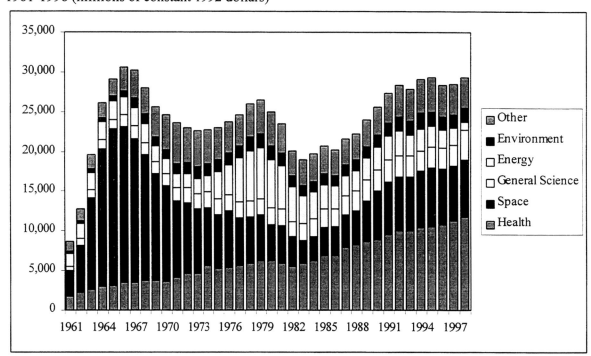

Notes: For fiscal years, 1961-1977, data are federal obligations. For fiscal years, 1978-1998, data are for budget authority. Beginning in fiscal year 1998 a number of Department of Energy programs were reclassified from energy to general science.

Source: NSF 1999d, Table 25.

decreased slightly from 10 percent in 1970 to 7 percent in 1998. Meanwhile, institutional funds increased from 11 percent of total academic R&D funding in 1970 to more than 19 percent in 1998. Industrial support has grown faster than other sources of academic R&D funding, increasing from less than 3 percent in 1970 to 7 percent in 1990, where it has remained (NSF 1998d). Since the federal government nevertheless remains the largest funder for academically-performed R&D, changing federal funding levels and priorities affects college and university research and development activities. Alan Rapoport has recently demonstrated that between 1973 and 1996, field shares of total academically-performed R&D have increased for the medical sciences, engineering, computer sciences, and astronomy; remained roughly constant for the mathematical sciences; and

declined for the social sciences, psychology, environmental sciences, physics, chemistry, biological sciences, and agricultural sciences. These changes are due, in part, to shifts in federal funding streams noted above (NSF 1998b).

The National Academies' Committee on Science, Engineering, and Public Policy (COSEPUP) has observed that recent federal obligations for academic R&D may continue these trends in funding by field and warns that it is important to invest "in a balanced way across a broad range of fields to maintain the overall health of the science and technology portfolio." COSEPUP notes that federal obligations for academic R&D would increase by 16.6 percent in constant dollars from FY 1997 to FY 2000 under the President's budget proposal. However, when funding from the

Department of Health and Human Services (99 percent of which is from the National Institutes of Health) is excluded, support for academic R&D would increase only 1.9 percent in constant dollars from FY 1997 to FY 2000. Moreover, DOD support for academic R&D, which funds a large proportion of R&D in engineering, material science, computer science, and mathematics, would decrease 31.7 percent in constant dollars from FY 1997 to FY 2000 (NAS 1999).

Data For Research and Innovation Policy

Investment in research and development is commonly thought to provide high rates of return to the economy and that subsidizing it is an appropriate activity of the federal government. For example, the House Science Committee report *Unlocking the Future* recently argued "because the scientific enterprise is a critical driver of the Nation's economy, investment in basic scientific research is a long-term economic imperative" (U.S. House of Representatives 1998). The Progressive Policy Institute (PPI) also recently asserted that in addition to increasing venture capital investment and growing R&D expenditures by firms, federal support for R&D remains important and "has significant economic payoffs" (Atkinson and Court 1998). Similarly, the Council on Competitiveness argues that "now, more than ever, research and development (R&D) drives the process of innovation that underpins our nation's economic well-being and national security" (Council on Competitiveness 1998). The Committee on Economic Development began its recent report on basic research by arguing that "basic research is a critically important—yet often undervalued—source of American economic growth and prosperity (Committee on Economic Development 1998). Economists, too, note that the outcomes of scientific research lead to economic growth.

"The lags between basic research and its economic consequences may be long," Paula Stephan writes, "but the economic impact is indisputable" (Stephan 1996). David Mowery writes that several factors have aided innovation in the U.S. economy and the competitive position of U.S. firms. One of these is the U.S. domestic market which "remains that largest high-income region that possesses unified markets for goods, capital, technology, and labor." Another is a federal policy structure that supports economic experimentation through appropriate competition policies, intellectual property rights, and "large-scale federal funding of R&D in universities and industry" (Mowery 1999).

While government investment in R&D is held to be positive, how the government invests funds in R&D is an important policy issue that requires relevant data for proper analysis. Changes in the structure of the U.S. R&D system discussed above have led to new questions about the allocation of these resources in the 1990s:

- What are appropriate sources and levels of short- and long-term R&D support?
- What should the level of direct federal support for R&D be, and how should it be allocated across agency missions, across fields of science and engineering, and across short- and long-term research activities?
- How have shifts in federal priorities (i.e., decreases in defense R&D; increases in health research) affected the balance of funding among fields?
- What is the role of multidisciplinary research or integrative research across fields in the overall science and technology portfolio?
- What is the nature of R&D and innovation in the service sector and what does this mean for the federal investment in science and technology given its role in promoting economic growth?

- What are the future roles of federal regulatory, technology, trade, and macroeconomic policies in supporting industrial innovation?
- What role should strategic alliances among firms, universities, and government play in the performance of R&D?
- Do universities have adequate facilities for research in light of changing federal priorities and of the increase in strategic alliances among industry, government, and academia?
- What do the existence of geographic clusters of innovation or the globalization of R&D suggest for the way federal funds for R&D are allocated?

In the remainder of this chapter we examine specific changes SRS should make to improve its data on R&D funding and performance in order to help answer these questions.

Industrial R&D Statistics

The principal source of national data on industrial R&D is the Survey of Industrial Research and Development (RD-1) conducted for SRS by the Census Bureau. Changes in the U.S. R&D system since the 1980s have already had implications for data collected by SRS on the funding and performance of R&D by private firms through this instrument. From its inception through the 1970s, RD-1 served as a suitable instrument for collecting industrial R&D data and NSF changed the instrument little during these decades. In the 1980s, however, as the economy and industrial R&D began to change, the instrument became increasingly non-reflective of current R&D organization and practices:

> In the 1980s, issues of U.S. industry's competitiveness, rapidly changing technology, and globalization interrupted the survey's stability and challenged its design and objectives. These issues first became evident in the R&D survey with the changing dynamics of company

organization. Simply maintaining the survey panel in the midst of countless mergers, acquisitions and divestitures was a demanding process (Champion 1998).

Other issues became evident: the increase in R&D performed by non-manufacturing firms; a growing number of small, start-up companies conducting R&D; and an apparent increase in the number of intra- and inter-sectoral R&D partnerships.

In the 1990s, SRS substantially modified the sample design for RD-1 to begin addressing some of the issues raised by this changing world. First, evidence of increased R&D in the service sector compelled SRS to draw a sample that included many more non-manufacturing firms from a broader range of non-manufacturing industries. Second, the approach SRS was using to administer RD-1 and estimate industrial R&D was missing R&D performed by a variety of firms. For example, SRS drew a sample of 14,000 firms in 1987 and administered RD-1 to them in that year. From 1988 to 1991, however, only about 1,700 of these firms were annually surveyed; data for the other 12,300 or so firms was estimated. Among other things, this approach was inadequate for capturing R&D performed by smaller firms and by firms that were new performers of R&D. To provide more reliable estimates of R&D, therefore, SRS not only expanded the representation of non-manufacturing industries and small companies, but also decided to draw an expanded sample of 25,000 firms annually (NSF 1999g).

SRS also added questions to RD-1 on company-sponsored R&D in addition to company-performed R&D. This was initially added as a quality control measure, but it consequently provided meaningful information on outsourced R&D and R&D performed by foreign subsidiaries. The Census determined that companies could report foreign R&D by country and added this question to the survey in 1993 (Champion 1998).

Still, a number of issues remain. First, given the large number of multi-product firms, should RD-1 be fielded at the firm level, as is presently done, or should it be fielded at the line-of business level? Second, since R&D performed in the service sector appears to differ substantively from R&D in the manufacturing sector, are we collecting the right data on one-quarter of industrial R&D? Third, because translating scientific and technological advances into new processes and products is a critical component of generating returns on R&D investment, what useful data and analyses could be obtained on industrial innovation to inform policymakers, managers, and researchers? Fourth, how does an apparent increase in the number of intra- and inter-sectoral partnerships affect how R&D data should be collected?

Improving Data on R&D by Industry Group

At present, the long version of RD-1 asks firms to provide data every other year on how their expenditures for applied research and for development break down by product group. However, high nonresponse to this question leaves the resulting data of questionable value. Many firms have difficulty categorizing applied research and development by these fields, or if they can categorize them, they do not want to disclose them because the data are considered strategic or proprietary. For these reasons, nonresponse to product field for applied research and development is unlikely to improve. We recommend that SRS drop this question from the survey.

Although SRS may be constrained in this area because it has very limited authority to compel responses from surveyed parties, we believe SRS should investigate options available for obtaining more detailed and accurate data on industrial R&D expenditures

in lieu of the product group question. Currently, Census administers RD-1 to firms. All R&D spending within a firm is attributed to the firm's industry classification. Thus, if 51 percent of a firm's business is in motor vehicles and 49 percent in other products, all of its R&D expenditures are counted as in "motor vehicles." As a result, such data collection likely skews how industrial R&D is portrayed by industrial classification.

Administering RD-1 to business units rather than firms as an option for obtaining more accurate R&D data by industrial classification was the subject of much discussion at the Workshop on Research and Innovation Indicators held by the NRC's Board on Science, Technology, and Economic Policy (STEP) in February 1997. The report summarizing the workshop's findings maintains that SRS should collect data by administering RD-1 to business units:

> In an economy in which large, complex, multiproduct firms perform most R&D activities, the business unit represents a more homogeneous set of activities, whereas combining responses on a range of variables for a variety of products and processes will obscure significant industry-specific conditions that affect technological innovation. Moreover, managers at the business unit level are likely to be better informed about innovation-related investments and performance measures than are corporate headquarters officials (NRC 1997c).

Workshop participants concluded that RD-1 should be administered to business units, and should collect data on "R&D expenditures, composition of R&D (process versus product; basic research, applied research, and development), share of R&D that is self-financed, supported by government or other contract, as well as contextual information on business unit sales, domestic and foreign, and growth history of the business unit" (NRC 1997c).

SRS should carefully consider a variety of questions in assessing the costs and benefits of administering RD-1 to business units. For example, it should investigate whether the business unit, perhaps well enough understood by business managers, is sufficiently well defined in operational terms to be a useful construct with which it could collect R&D data reliably. SRS should also examine and learn from the experiences of the Federal Trade Commission in the late 1970s in the administration of business unit surveys. Collecting R&D data at the business unit level would allow more accurate detail on R&D by product or service. In addition, obtaining the geographic location of the business unit would allow these R&D data to be linked with other locational data, such as demographic, educational, and economic data by metropolitan statistical area. Of considerable importance, though, is whether data collected at the business unit level could be aggregated to the firm level, since other economic data to which RD-1 might be linked are available at that level.

R&D and Innovation

David Mowery argues that current R&D expenditure data do not provide adequate information on many activities contributing to innovation such as "investments in human resources and training, the hiring of consultants or specialized providers of technology-intensive services, and the reorganization of business processes." He also contends that R&D expenditure data "do not shed much light on the importance or content of the activities and investments essential to inter-sectoral flow and adoption of information technology-based innovations." He notes further that in such non-manufacturing industries as trucking or food retailing R&D inputs may not be adequately captured because they are indistinguishable from other corporate expenses on operations, materials, or marketing. These contentions are of particular concern, because the second

structural change in industrial R&D in the last two decades has been the increasing share of R&D that is performed within non-manufacturing firms. SRS and Census have expanded the sample for RD-1 to include more non-manufacturing firms, and this has improved the data available on formal R&D funding. However, these data may not fully capture aspects of how R&D is performed and how its outputs are adopted. "Without substantial change in the content and coverage of data collection," Mowery advises, "our portrait of innovative activity in the U.S. economy is likely to become less and less accurate" (Mowery 1999).

In light of these problems, SRS should collect additional data or conduct different kinds of analyses to better understand innovation in the service sector as well as in manufacturing. First, SRS should pursue plans to develop a survey of industrial innovation as recommended in the report of the 1997 STEP workshop as one means for providing data on technology adoption and the development of new processes and products (NRC 1997c). SRS should examine and learn from the experiences of the many individuals and organizations who have attempted to administer innovation surveys, from the efforts of Hansen, Hill, Stein, and More in the early 1980s to its own pilot survey in 1994 (Hill, Hansen, and Stein 1983, Hansen, Stein, More 1984). As part of this review SRS should examine innovation surveys conducted in other countries. SRS should include both potential survey respondents and data users in determining the kinds of research questions that would be addressed by such a survey, in developing the survey instrument, and in determining appropriate means for administering the survey.

Second, SRS should also conduct or sponsor further research into the nature of service sector R&D. For example, since technology is as often diffused by the movement and interaction of scientists and engineers as by other means, SRS could examine whether human resources data can be used to explore important characteristics and trends in research utilization and industrial innovation in both the service and manufacturing sectors. As one individual interviewed for this study noted, analysts first understood that the service sector was an increasingly important locus of R&D performance when scientists and engineers moved into the financial services sector. Human resources data could potentially shed light on technological diffusions between sectors and fields as seen in career mobility, the nature and extent of intra- and inter-sectoral partnerships and alliances, and the locational aspects of R&D and innovation.

An examination of how human resources data could be used to characterize research utilization and industrial innovation would also provide an excellent opportunity for SRS to examine how data sets could be linked to improve analytical range and depth, as we also recommend elsewhere. In this case, linking SRS human resources data with data on patents and publications, R&D funding and performance data, and other demographic and economic data available by geographic location could potentially enrich analysis.

Partnerships

"Beginning in the 1980s, a combination of severe competitive pressure, disappointment with perceived returns on their rapidly expanding investments in internal R&D, and a change in federal antitrust policy led many U.S. companies to externalize a portion of their R&D," writes David Mowery (Mowery 1999). Indeed, indicators do point to increasing numbers of joint ventures, strategic alliances, and cooperative research agreements:

- From 1985 to 1996, more than 665 research joint ventures (RJVs) among firms working on generic, pre-competitive research have been registered with the

Department of Justice under the National Cooperative Research Act (NSB 1998a).

- The 1980s and 1990s saw an increase in the number of partnerships formed among high-technology firms (information technology, biotechnology, and new materials) for technology development or technology transfer. Over time, research joint ventures have been replaced by contractual and equity arrangements as the primary vehicles for arranging strategic technology partnerships. International partnerships have increased but not at the same rate as domestic ones (Hagedoorn 1996).

- The fastest growing source of funding for academically-performed R&D is industry, which financed less than 3 percent in 1970, but increased its share thereafter to 7 percent in 1990, where it has remained (NSF 1998d).

- 3,512 Cooperative Research and Development Agreements (CRADAs) were executed between federal laboratories and private firms from 1992 to 1995 (NSB 1998a).

These indicators are suggestive of an important trend, and even of structural change, but they do not provide a complete picture of the role and variety of partnerships. As the STEP report on *Securing America's Industrial Resurgence* points out, "the incidence and value to firms of outsourcing R&D are unclear" (NRC 1999). Changes to SRS survey instruments could help illuminate these phenomena further.

SRS currently asks respondents to the Survey of Industrial Research and Development to provide the amount of federal funds, company funds, and total funds spent for research and development performed in the United States on behalf of the firm by others outside the company. The Survey also asks for an aggregate of company funds for research and development performed by foreign subsidiaries or other organizations outside the United States. To further describe the nature of partnerships, SRS should examine the efficacy of asking firms to provide data on the number of new, continuing, and terminating alliances each year and how this might vary by type of alliance. Data that Hansen, Hill, and Stein collected in their 1981 and 1983 surveys would be a useful reference in regard to response rates, definitional problems, and other aspects of collecting this sort of data (Hill, Hansen, and Stein 1983). SRS might also ask firms to break out their responses to R&D performed with company funds overseas into that performed by foreign subsidiaries and that performed by foreign subcontractors.

At our September workshop, Patrick Windham noted that some large firms, such as Cisco Systems, acquire start-up companies as an R&D strategy. Other firms gain access to R&D and innovations by making equity investments in other companies. These practices are thought by some to be increasingly common in high-technology industries, such as information technology and telecommunications. If they are widespread, then an accurate picture of the total amount of corporate R&D spending must account for these kinds of investments. SRS should work with experts in this area to investigate how widespread this practice is, and how it may be characterized in certain industries, especially in the information technology industry. Based on the results of its investigation, SRS should consider whether to account for these phenomena in the Survey of Industrial Research and Development or by other means.

Some analysts have suggested that industry has compensated for the downsizing of central labs by contracting research to university- or nonprofit-based researchers. The decentralized nature of some large corporate R&D operations makes it difficult for firms to track these alliances effectively. Even if they could, firms might balk at

providing data on outsourcing that would expose corporate R&D strategies. In 1992 and 1993, the Census Bureau interviewed companies about the feasibility of reporting several new data items on the Survey of Industrial R&D, including the type of organizations that receive subcontracts. For the proposal, Census developed a list of organization types that might receive funds from industry to perform R&D. Census found that their list was inadequate. However, they did not revise it because respondents indicated they would have considerable difficulty in providing the information because grants could be made to universities outside of the R&D budget, and it may be difficult to distinguish between research and other awards (Champion 1998).

This remains an important issue, however, and several alternative means for collecting these data may be identified. For example, SRS might explore whether detailed data on the extent of industry-university partnerships could be collected through the Survey of Research and Development Expenditures at Universities and Colleges. Currently, this survey asks for the aggregate amount of R&D funding the institution received from industry, but it could also ask for this amount to be disaggregated by field as it does for total expenditures and federally funded expenditures, thereby providing more detail in this area. SRS could also ask academically-based respondents to the Survey of Doctorate Recipients about the number and type of relationships they have with industry and the dollar amounts associated with them.

Allocating Federal Funds for Research and Development

In the face of constrained budget resources and changing federal priorities for scientific and technological research, the U.S. Congress called on the National Academies in 1994 for guidance in developing a more systematic and prioritized approach to federal R&D funding. The Senate Appropriations Committee that year requested a study from the National Academy of Sciences, National Academy of Engineering, and Institute of Medicine to address:

> the criteria that should be used in judging the appropriate allocation of funds to research and development activities, the appropriate balance among different types of institutions that conduct such research, and the means of assuring continued objectivity in the allocation process.

The resulting NRC report, *Allocating Federal Funds for Science and Technology*, recommended a new budget process for science and technology to ensure that the United States continues as the world leader in these areas by funding agency missions and the broad range of science engineering fields. Data on resources for the science and engineering enterprise should inform the process for allocating federal research investments and the examination of the federal science and technology budget (NRC 1995).

Priority Setting in Federal Science and Technology

Allocating Federal Funds laid the groundwork for subsequent debate about resource allocation by focusing on the "federal science and technology" (FS&T) budget. The FS&T budget "would reflect real investment in the creation of new knowledge and technologies and exclude activities not involving the creation of new knowledge or technologies, such as the testing and evaluation of new weapons systems." The FS&T budget includes funding for basic and applied research from all departments and agencies and all civilian development funding. However, it only includes that part of defense development at the Department of Defense (DOD) and the Department of Energy (DOE)

that includes generic technology development (6.3 at DOD and its equivalent in the DOE atomic energy defense program). For FY 2000, federal research and development in the President's proposed budget is $78.2 billion while the FS&T budget, a subset of the former, would be $49.6 billion (NAS 1999).

Among other goals, the report wanted to assure that science was funded adequately across field, recommending that "departments and agencies should make FS&T allocation decisions based on clearly articulated criteria that are congruent with those used by the Executive Office of the President and by Congress" and "the President and Congress should ensure that the FS&T budget is sufficient to allow the United States to achieve preeminence in a select number of fields and to perform at a world-class level in the other major fields" (NRC 1995).

Four years later, the budget context for research appropriations has not changed dramatically, even with the advent of a budget surplus, since caps on discretionary spending remain in place. As a result, calls for prioritization in the allocation of resources for research continue. The Academies continue to follow this issue by providing annual observations on the FS&T budget and by developing international benchmarks by field (NAS 1999). In presenting the "21st Century Research Fund" as a component of the President's fiscal year 2000 budget submission, Neal Lane, the President's science advisor, noted that the Fund was designed to reflect the National Academies' FS&T concept. Indeed, the President's budget request states "This budget also reflects an effort to re-establish an optimum balance between health care research and other scientific disciplines—a concern voiced in recent years throughout the science community" (U.S.O.M.B. 1999). *Science* magazine recently took an editorial position in favor of a more "scientific" approach to establishing priorities for allocating federal funds for science research (Bloom 1998). This issue was also raised by interviewees as a critical resources issue for the post-Cold War era that has yet to be resolved.

The National Science Board has also taken an active role in priority setting across agencies and fields. The Board recently recognized in its *Strategic Plan* that the Academies' report on *Allocating Federal Funds* laid "the foundation for future efforts in what remains a formidable challenge." Still it has noted that "presently, there is no widely accepted way for the federal government, in conjunction with the scientific community, to make priority decisions about the allocation of resources in and across scientific disciplines." The Board has argued that this is true both of the OMB budget process and the congressional appropriations process. While recognizing that this is a "difficult and controversial" task, the NSB has determined to "conduct a state of the art assessment for methodologies for priority setting for research, including an examination of the experiences of other countries" (NSB 1998d).

SRS should continue to monitor whether the Survey of Federal Funds for Research and Development provides data that policymakers, researchers, and others require to measure the concepts in funding for fields of science and engineering they find important at this time.

First, at a recent SRS-sponsored Workshop on Federal Research and Development, participants suggested a more accurate portrayal and analysis of the FS&T could be obtained if data provided by DOE and NASA more explicitly specified the science and technology components of their research and development budgets. The Department of Defense (DOD), for example, does this by specifying spending for basic research, applied research, and advanced technology as distinct from weapons systems development in its research and development program. Also, FS&T data that can now be accessed are available only at aggregate funding levels. Disaggregating federal R&D funding into FS&T and non-FS&T components for more detailed trends—such as trends in federal support for research at U.S. colleges and universities—would further enable more meaningful analysis.

Second, SRS should also continue to work toward resolving discrepancies in the results obtained by the Survey of Federal Funds for Research and Development, the Survey of Industrial Research and Development, and the Survey of Research and Development Expenditures at Universities and Colleges. As shown in Figure 5-5, the discrepancy in the level of federal obligations to colleges and universities obtained by the federal funds survey and the level of federally-funded R&D expenditures reported by colleges and universities in the academic R&D survey has grown since 1992. By 1997, the difference in results between the two surveys was more than $1.9 billion. As noted in Chapter 3, SRS is aware of this discrepancy, has begun investigating it, and has some preliminary observations about the reasons for the discrepancy (e.g., increasing research subcontracting among universities that results in double counting, differences between federal obligations and expenditures, etc.)

If these data sets are to continue to have credibility and usefulness as sources of data for analysis of the distribution of federal funding across fields, they should produce similar trends by field even if the aggregate totals differ. Table 5-1 displays data by field on federal obligations for research to colleges and universities (as reported through the Survey of Federal Funds for Research and

Figure 5-5 SRS Data on Federal Obligations to Universities and Colleges for Research Compared to SRS Data on Federally-Funded Academic Research and Development Expenditures, 1971-1997

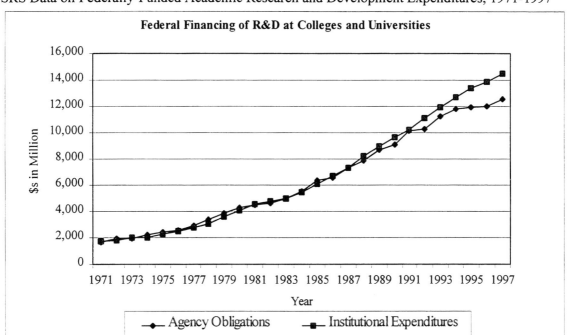

Source: Obligations: NSF/SRS, Survey of Federal Funds for Research and Development; Expenditures: NSF/SRS, Survey of Research and Development Expenditures at Universities and Colleges.

Development) and federally-funded R&D expenditures by colleges and universities (as reported through the Survey of R&D Expenditures by Universities and Colleges).

The data show very different funding trends by field. For example, federal obligations for academic electrical engineering research decreased in constant dollars from 1993 to 1997 by 31.6 percent. At the same time, according to the academic R&D survey, federally-funded R&D expenditures in that field increased by 27.2 percent. The Federal Funds survey provides data by field for research categories only and the academic R&D survey provides data by field only for research and development combined. Thus, this table is, in a sense, comparing "apples" with "apples and oranges." However, development is only about ten percent of federally-funded academic R&D, so it is hard

to imagine that this by itself would account for such different results. For these data to be useful for analysis of funding by field these discrepancies must be accounted for.

Third, SRS should continue to work with agencies to ensure that data reported by field is accurate and consistent over time. Interviews conducted for this study suggested that agency respondents work with contract classification systems that may have evolved long ago and may no longer fit current classifications of fields in science and engineering. In addition, the budget officer who is the respondent to the federal funds survey may base the field classification on the title of particular procurement in a very subjective way. SRS has begun working with policymakers in agencies that use these data to encourage them to impress on those who reply to the SRS survey the importance of providing good,

Table 5-1 Federal Obligations to Universities and Colleges for Research Compared to Federally-Funded Research and Development Expenditures, by Field, 1993 and 1997

	Federal Obligations for Academic Research			Federally-Funded Academic R&D		
	Constant 1999 dollars		% Change	Constant 1999 dollars		% Change
	1993	1997	1993-1997	1993	1997	1993-1997
S&E, total	10,659	11,050	3.7%	13,352	14,866	11.3%
Engineering, total	992	1,013	2.1%	2,075	2,326	12.1%
Aeronautical/Astronautical	82	69	-15.8%	179	185	3.3%
Chemical	73	65	-11.0%	160	171	7.2%
Civil	42	46	8.7%	172	202	17.4%
Electrical	219	150	-31.6%	512	651	27.2%
Mechanical	131	78	-40.4%	346	335	-3.2%
Metallurgical/materials	225	268	19.2%	168	229	36.4%
Other, n.e.c.	222	338	52.2%	538	553	2.6%
All Sciences, total	9,667	10,037	3.8%	11,277	12,540	11.2%
Physical sciences	1,310	1,193	-8.9%	1,690	1,748	3.4%
Astronomy	134	172	28.5%	184	190	2.9%
Chemistry	389	350	-10.1%	564	572	1.4%
Physics	679	535	-21.2%	791	836	5.8%
Other, n.e.c	108	136	25.8%	150	150	0.0%
Environmental sciences	659	689	4.5%	971	1,060	9.2%
Atmospheric	179	211	18.2%	180	191	6.0%
Earth Sciences	211	127	-39.8%	271	278	2.4%
Oceanography	164	220	34.2%	367	385	4.9%
Other, n.e.c.	107	130	21.4%	152	205	35.0%
Mathematical sciences	152	121	-20.4%	227	211	-6.9%
Computer sciences	395	464	17.4%	472	527	11.5%
Life sciences	6,143	6,731	9.6%	7,133	8,079	13.3%
Agricultural sciences	166	177	6.6%	503	582	15.8%
Biological sciences	3,246	3,739	15.2%	2,580	2,786	8.0%
Medical sciences	2,609	2,601	-0.3%	3,764	4,400	16.9%
Other, n.e.c.	122	213	75.1%	287	310	7.8%
Psychology	321	290	-9.5%	261	276	5.5%
Social sciences	241	202	-16.3%	378	428	13.5%
Other sciences, n.e.c.	446	346	-22.3%	144	211	46.6%

Sources: NSF/Survey of Research and Development Expenditures at Universities and Colleges (NSF 1999a, Table B-4); and NSF/Survey of Federal Funds for Research and Development (NSF 1999c, Table 62).

consistent responses. We encourage SRS to continue with this work. At the same time, agencies must be very careful about how they do make changes when they are warranted. For example, a recent analysis using data from the Federal Funds Survey found that it was difficult to track changes by fine field over time because the National Science Foundation, in responding to the survey, had changed its procedures for classifying research obligations by field beginning in fiscal year 1996. This resulted, for example, in substantial increases that year in such areas as "engineering, n.e.c." and "physics, n.e.c."[2] (NRC 1999).

Fourth, as we have recommended elsewhere, SRS should use a common science and engineering taxonomy across its surveys to facilitate linkages across its data sets—and especially between its R&D investment and human resources data sets—in order to better support analysis of the allocation of science and engineering resources. In a recent analysis of trends in federal spending on science and engineering research, Michael McGeary and Stephen Merrill sought to examine how such trends have affected funding for specific science and engineering fields, and moreover, graduate training in those fields. While they were able to look at trends in federal support for graduate training in a very broad way, the differences in field taxonomy for the Federal Funds Survey and the Survey of Graduate Students and Postdoctorates in Science and Engineering made it difficult to associate changes in federal research by field with changes in graduate training by fine field (NRC 1999, Appendix A). The ability to link data sets would allow such analyses to be carried out in a more detailed and effective manner.

Interdisciplinary Research

SRS should also investigate further the degree to which inter- or multidisciplinary research is increasingly prevalent and important in scientific and technological research in the United States. "Interdisciplinary research is a mantra of science policy," write Norman Metzger and Richard Zare. "Virtually any meeting on the current state and future of science is leavened by obligatory statements about the importance of enabling researchers to work seamlessly

across disciplinary boundaries and by solemn declarations that some of the most exciting problems in contemporary research span the disciplines." Metzger and Zare argue, however, that the way federal agencies currently fund research does not encourage as much interdisciplinary research as there ought to be since such research would "enrich and enlarge" science and engineering in the United States by reintegrating knowledge (Metzger and Zare, 1999).

Data on the extent of interdisciplinary research conducted are important to analyses of science and engineering resources for two reasons: they are critical for accurately measuring the science and engineering enterprise generally, and they have implications for priority setting by field within that enterprise if certain projects are indeed inter- or multidisciplinary. Yet how much federally-funded R&D is multidisciplinary in nature is not clearly known, partly due to the limitations of SRS data. Data from NSF's Federal Funds survey or other data sources do not reveal the extent or nature of multidisciplinary research supported by the federal government, industry, or other organizations at present. Only one federal agency, the U.S. Department of Agriculture, is required by law to track how much of its research is multidisciplinary.

SRS has taken some preliminary steps to explore the nature and extent of inter-disciplinary research, but more needs to be done. For example, SRS explored the recent increases within major fields of federal funding that is "not elsewhere classified" (i.e., categorized by fine field) on the theory that much multidisciplinary research may be so classified. However, other kinds of projects or research in newly emerging fields (e.g., bioengineering) may also be classified as n.e.c. until the taxonomy is changed to provide its own category. Thus, n.e.c. is not a precise proxy for multidisciplinary work. Several participants at the SRS Workshop on Federal Research and Development suggested that

---

[2] Agencies report funding for projects that cannot be classified in a fine field, such as mechanical engineering, as "not elsewhere classified" (n.e.c.), such as in "engineering, n.e.c."

other agencies, in addition to USDA, should explicitly track how much of their funding is for multidisciplinary research, but others believed it would be difficult to sort projects in this manner, and that such a procedure would add to the respondent burden. That latter claim should be investigated further by SRS.

SRS should continue to investigate how the extent of multidisciplinary research may be obtained through surveys in both its R&D and human resources statistics programs. The division should consider holding a workshop or commissioning a study to better understand the nature and extent of this phenomenon and how to implement changes consistently throughout its data collection activities, as needed and where possible. The workshop or study should provide guidance or recommendations on how to consistently implement such changes.

Federal Funding of Academic Facilities

Concerned about the condition of research facilities at U.S. colleges and universities, the U.S. Congress mandated that NSF collect data on academic science and engineering (S&E) research facilities issues. These issues include the amount and condition of current space for conducting academic S&E research, the percentage of existing space that needs to be repaired or renovated, the need for new construction of research space, the extent of current repair work and construction, and the sources of funds for repair and new construction (NSF 1999j).

The federal government bears only a small share of the costs of building and maintaining academic research facilities in the United States, but the federal government has struggled throughout the 1990s with ensuring that reimbursements to major research universities for facilities and administrative (indirect) costs are proper. In 1990 and 1991,

institutions used their own resources to fund 55 percent of new construction, renovation, and repair, while state and local government provided 32 percent, and the federal government 14 percent. In 1996 and 1997, the respective shares of funding from these sources were 62, 29, and 9 percent. Nonetheless, the U.S. Office of Management and Budget (OMB) sought in 1998 to revise OMB Circular A-21 on university indirect cost reimbursement for facilities. Through the revised circular, OMB instructed institutions to document—using data from SRS's Survey of Scientific and Engineering Research Facilities at Universities and Colleges and other sources—their construction costs for new buildings so that they could be compared with similar costs for other institutions. Associations for the universities, however, argued that existing data from that NSF survey were not adequate for comparing costs among institutions.

SRS should continue to cooperate with OMB in the development and refinement of a survey that would follow up the Survey of Scientific and Engineering Research Facilities at Universities and Colleges to collect detailed data on the space and cost of buildings at academic institutions with science and engineering facilities with construction costs over $25 million. This follow-up questionnaire should collect data for analyzing reimbursement for federally funded research facilities through the application of the facilities and administrative (indirect cost) rate. The follow-up survey, focusing on buildings, the use of which is anticipated to be over 50 percent for research, should assess the cost of buildings by science and engineering discipline. The survey should meet OMB's needs and also the needs of the National Center for Education Statistics (NCES) for use of square footage. We believe that this is an important test of whether SRS data can directly support an important federal science and engineering policy question. SRS should cooperate with OMB in this endeavor and revise the Facilities survey to meet this need. Another question that may be addressed by the Survey of Scientific and Engineering Research Facilities at Universities and Colleges concerns the financial requirements for renewing the science and engineering infrastructure across all fields at academic institutions. It should examine the feasibility of posing a one-time question to obtain an estimate of the costs associated with future renovation of the academic research infrastructure. Data on these requirements, especially in light of ongoing concerns about federal reimbursement of indirect costs, would help plan for an orderly renewal of academic infrastructure.

6

Recommendations

NSF's Division of Science Resources Studies is a federal statistical agency that exists to serve the information needs of policymakers, managers, educators, and researchers in the science and engineering community. The division's mandate is "to provide a central clearinghouse for the collection, interpretation, and analysis of data on scientific and engineering resources, and to provide a source of information for policy formulation by other agencies of the Federal government" as required by the National Science Foundation Act of 1950, as amended.

To be effective in this role, SRS must ensure the ongoing relevance of the information it provides through its portfolio of data collection and analysis activities. In this chapter we summarize our recommendations for operational changes that will facilitate an ongoing renewal of the concepts that SRS data should measure and the analysis the division provides. We also summarize the issues in science and engineering resources that we recommend SRS should better address through its data collection and analysis activities. We have assigned our highest priority to our first

two recommendations that address how SRS may strengthen its dialogue and interactions with both its stakeholders and external researchers. These are followed by recommendations that provide an immediate agenda for the strengthened dialogue and interaction.

Ensuring Relevance and Establishing Priorities

Relevance of data has many dimensions and each of these should be considered in determining whether a statistical agency is meeting the needs of its constituents. (NRC 1997a) For purposes of this study, we have focused on three factors that affect the relevance and analytic value of data: appropriateness of concepts and their measurement; ability to link data sets; and data currency. Here, we examine operational aspects of a federal statistical agency that affect these dimensions of relevance and the quality of the data and information SRS provides to its constituents.

Appropriateness of Concepts and their Measurement

Recommendation 1. To keep its data relevant and maintain data quality and analytic capacity, SRS should adopt a strategy of continuous review and renewal of the concepts it seeks to measure, and revise its survey instruments, survey operations, and data analysis as needed to keep them current. To achieve this, SRS must strengthen the frequency and intensity of its dialogue and interactions with data users, policymakers, and academic researchers and develop internal processes to convert the feedback it receives from these stakeholders into changes in its surveys and analyses. A key element of this strategy is the creation of advisory committees for SRS surveys that would assist SRS in establishing priorities for future change.

To expand the range of surveys that benefit from advisory committees, we strongly recommend the creation of such a committee for the Survey of Industrial Research and Development. We also recommend that the existing Special Emphasis Panel (i.e., advisory committee) for the Doctorate Data Project (DDP) advise SRS on the Survey of Earned Doctorates (SED), the content of all three SRS personnel surveys, and the design of the Scientists and Engineers Statistical (SESTAT) system. This panel already provides SRS advice on the SED and the Survey of Doctorate Recipients (SDR), and should also provide advice on the National Survey of College Graduates (NSCG) and the National Survey of Recent College Graduates (NSRCG).

A statistical agency should work with data users to define the concepts that it will measure to meet users' information needs. These concepts and how to measure them should be continuously reviewed and revised as issues change and as analysis reveals alternate measures that better capture information that is useful to constituents.

Special attention should be paid to whether new concepts can be quantified in a meaningful way, whether data may be reliably collected on the subject, the level of detail that meets user needs, and the cost-effectiveness of data collection.

To operationalize this process of ongoing review and renewal of data concepts, SRS must increase the frequency and intensity of its dialogue and interactions with data users, policymakers, and academic researchers to capitalize on their insights, expertise, and analytic capabilities. To generate direct interaction of this sort, SRS should establish an advisory committee for each of its surveys. These committees can assist SRS in keeping survey content up-to-date and also in establishing priorities for future change.

SRS already has such committees (also referred to as "special emphasis panels") for many of its surveys. We urge that an advisory committee be instituted for the Survey of Industrial R&D given changes now occurring in industrial R&D. We also urge that the Special Emphasis Panel for the Doctorate Data Project (DDP) be engaged for advice on the Survey of Earned Doctorates and on all three personnel surveys in the SESTAT system. The 1989 NRC Report, *Surveying the Nation's Scientists and Engineers*, recommended that an advisory committee be established for the SESTAT system to review its design and content for the next decade. We believe that expanding the scope of the DDP panel is the most efficient means for accomplishing this.

The SRS Breakout Group of the Directorate for Social, Behavioral, and Economic Sciences (SBE) Advisory Committee that provides NSF with advice on SRS surveys and operations and the advisory committees for individual surveys should work together to review and assist in the implementation of our recommendations, and set priorities among them and other proposals for change. To facilitate this interaction, the SRS Breakout Group of the SBE Advisory

101

Committee should include individuals who are members of advisory committees for specific surveys. By having members of individual advisory committees on the SRS Breakout Group of the SBE Advisory Committee, individual survey advisory committees may contribute to the process of ensuring the overall relevance of the SRS portfolio and establishing priorities for change among surveys.

SRS may also generate dialogue and interaction through other means. These include holding an ongoing series of workshops on issues emerging in the science and engineering enterprise, and improving outreach with constituent groups through booths at conferences and similar activities. SRS should also engage in more purposeful dissemination of publications and enhanced customer service as means for promoting interaction with data users. Finally, SRS should continue to field a periodic customer survey.

SRS should convert the feedback it receives from these stakeholders into means for producing the information its constituents seek. SRS may revise survey instruments or add special modules to instruments to collect data for one survey cycle. In the past SRS has had a quick response panel for issues in industrial R&D. Reinstituting such panels would provide another means for collecting information, particularly on issues that may be currently important, but which do not necessarily signify structural change. Finally, SRS may employ or sponsor qualitative research as a complement to periodic surveys in order to obtain information more rapidly or comprehensively on poorly understood issues.

Data Analysis

Recommendation 2. SRS should more actively engage outside researchers in the analysis of SRS data on current science and engineering resources issues. This may be accomplished by allowing researchers to work at SRS as visiting fellows and by establishing an external grants program. SRS should also monitor and summarize research using its data.

Federal statistical agencies analyze their own data to provide substantive analysis of specific subject area issues and also to understand better the uses and limitations of the agency's data. SRS should carefully consider how it may best engage in and support research on science and engineering resources in the future. It should continue its in-house analytical activities, such as those in the Integrated Studies Program, and also actively engage outside experts in analytical activities using SRS data.

Because the division's staff is limited in size, it cannot have expertise in the full range of subject areas upon which it may be called for data and analysis. SRS and its constituents would therefore benefit from a more interactive relationship between SRS and external researchers who focus on science and engineering resources issues. Programs to bring external researchers into SRS through a visiting fellows program or to provide grants to researchers who utilize SRS data would expand the analytical range of the division and promote data use. These programs would be especially valuable if targeted toward data sets such as SESTAT that are underutilized, especially relative to their cost of production. They would also increase interaction between SRS and external data users, producing insights that would benefit SRS staff, researchers, and the relevance and quality of the data.

Also, SRS should monitor and summarize research conducted by others using SRS data, especially data from the NSCG, NSRCG, and the SDR, which are currently underutilized. This could be assigned to a contractor responsible for research. Other federal statistical agencies provide similar summaries of research based on data of use to social scientists. If SRS were to do this, too, such summaries would be an aid to researchers as well as a source of information for SRS in its role of advising policymakers.

Recommendation 3. SRS and the National Science Board (NSB) should develop a long-term plan for Science and Engineering Indicators *so that it is smaller, more policy focused, and less duplicative of other SRS publications to free SRS resources for other analytical activities.*

NSB and SRS should develop a long-term plan for restructuring *Science and Engineering Indicators*. Individuals interviewed for this study as well as science and technology policymakers recently interviewed by SRS following the publication of the last *Indicators* volume suggest at least three possible futures for *Indicators*. The first of these, of course, is maintaining the status quo. As currently conceived, *Indicators* provides a wealth of information on science and engineering resources in the United States, and increasingly, in an international framework. Both the NSB and SRS benefit from this arrangement: it provides the NSB a means for highlighting important science and engineering issues; it allows SRS to showcase its data in a high-profile report considered by many an essential reference for quantitative information by science and technology policymakers. The second is for the NSB to reduce the amount of policy analysis in the volume and concentrate on the data presented. Those with this perspective believe that the policy analysis presently in *Indicators* is not very useful, while the data are. The third is for the NSB to make the document more focused on policy issues and less on data. Individuals holding this point of view suggest that *Indicators* would have a greater impact if it were smaller, less redundant with other SRS publications, and offered policy insights built on important indicators.

We believe that *Science and Engineering Indicators* should be smaller and more policy-focused. *Indicators* would have more impact

on science and technology policy if it focused on bringing analysis to a small set of indicators on issues driving the future of the science and engineering enterprise. There should be a sharper division between the work of a policymaking body such as the National Science Board and the work of a federal statistical agency such as SRS. *Indicators* is redundant with other publications of SRS data which could be referenced in *Indicators* and also linked via hypertext when published on the Internet. Substantial SRS resources— especially staff resources—which are now devoted to the production of this volume, would be freed for other analytic activities if the report were refocused.

Data Comparability and Linkages

Recommendation 4. SRS should increase the analytic value of its data by improving comparability and linkages among its data sets, and between its data and data from other sources. Standardizing its science and engineering field taxonomies and other questions across survey instruments is a critical step in this process. Resolving discrepancies in results from different surveys is another.

The ability to link data collected through various instruments increases the breadth and depth of data, and thereby, the ability of analysts to use them to address current issues. SRS's portfolio of data collection activities has been established over the past half century as a number of individual surveys that provide information on specific pieces of the science and engineering enterprise. SRS has recently begun to manage its surveys as components of a more integrated data system. For example, SRS created SESTAT in the early 1990s in response to the 1989 NRC report, *Surveying the Nation's Scientists and Engineers*, that called for a more integrated science and engineering personnel data system. SRS has also created WebCASPAR, an integrated data source on higher education institutions.

To increase the analytic power of its data, SRS should find ways to integrate its data sets beyond SESTAT and WebCASPAR. First, SRS should continue to improve comparability in questions and response categories across surveys. Its surveys often ask similar questions, yet in different ways. This is clearly evident with regard to questions on field of science and engineering. SRS should develop a standardized science and engineering field taxonomy for its surveys, apply it to each survey instrument, and then keep it up-to-date on an ongoing basis. Second, SRS should continue to investigate discrepancies in survey results among its R&D funding and performance surveys and implement changes in survey instruments and operations to address and resolve them. The division should also consider establishing a committee to develop a design for a new, integrated R&D data system. Such a committee might be charged with devising a system analogous to the integrated SESTAT personnel data system that would also account for the apparent increase in the number of intra- and inter-sectoral R&D partnerships. Third, SRS must develop means for linking its R&D investment data and its human resources data.

SRS must, of course, follow appropriate guidelines for maintaining confidentiality of records as it engages in or facilitates the process of linking data sets internally and externally. However, within the bounds of legal requirements, SRS should make every effort to make its data sets available to researchers, including those who seek to link them to other data. SRS may look to other federal agencies, such as the Census Bureau's Center for Economic Research, as a model for achieving this.

SRS should also help coordinate the data gathering activities of others to improve data availability and comparability with its own data. For example, SRS should encourage standardization in data collection by professional associations and universities on the early job market and career experiences of new Ph.D.s. SRS should also continue to play a leading role in collecting, coordinating, and standardizing international science and engineering resources data.

Finally, SRS should seek effective ways to allow researchers to link its data to those from other data sources, public and private. For example, SRS could seek ways to link its graduate education data with data collected by the Educational Testing Service (e.g., GRE scores). SRS could also explore the ability to link its personnel data to other federal data on fellowships, traineeships, and research awards or to public and private data on patents and publications. SRS should also add an indicator for metropolitan statistical area to each of its R&D funding and performance surveys. This would allow aggregations of data for each survey by metropolitan area and would open these surveys to linkages with other economic, education, and demographic data collected at the metropolitan level.

Data Currency

Recommendation 5. SRS must substantially reduce the period of time between the reference date and data release date for each of its surveys to improve the relevance and usefulness of its data.

The currency of data depends on the periodicity of each survey and the timeliness with which data are released. Data that are collected biennially, for example, are expected to have a "shelf life" roughly twice that of annually-collected data and their relevance depends on their ability to be "current" during that shelf life. This currency, in turn, depends on the timeliness with which collected data are released for public use. Timeliness is measured, in the case of SRS, by the time that elapses between the reference date in each survey and the date at which survey data are released. The timeliness and currency of SRS data have been discussed by data users for some time. To improve the currency of its data, SRS must continue its recent efforts to substantially reduce the period of time between the reference date and data release date for each of its surveys. Means for accomplishing this goal include using incentives for timely response, increased use of the Internet for data collection, and early release of key indicators.

SRS as a Statistical Agency

Recommendation 6. SRS should be seen as a federal statistical agency and should be supported in its efforts to meet fully those standards set for federal statistical agencies for independence, professional staffing, data quality, and data analysis.

Recommendation 7. SRS's budget is substantially smaller than those of other federal statistical agencies and may need to be increased given the growing importance of its subject area and our recommendations for new processes, data collection activities, and additional studies. Any budgetary increase must be based on clearer information from SRS on its allocation of internal staff and financial resources across its surveys and other activities and on a clearer sense of priorities among current and future surveys and activities, as developed in coordination with advisory committees for its individual surveys and the SRS Working Advisory Group of the SBE Advisory Committee.

NSF should see SRS as a federal statistical agency. While SRS is small compared with other federal statistical agencies, its staff of about forty are called on to carry out each of the major functions of a federal statistical agency: data collection and acquisition, quality assurance, preparation of tabulations and public use data files, data analysis, publication of reports, and data and report dissemination.

NSF should support SRS in its efforts to meet fully those standards set for federal statistical agencies regarding independence, professional staffing, data quality, and data analysis. SRS should develop a staffing plan that allows it to improve staff skills and augment staff expertise, especially in key areas. First, SRS should continue to improve statistical and analytical skills among its staff through professional development activities conducted in-house or through the many courses offered in survey and statistical methods in the Washington, D.C. area. Second, SRS should augment its range of staff expertise in relevant subject areas through new hires. The number of full-time equivalent (FTE) positions in SRS was reduced earlier in the 1990s and has remained constant since. We recommend that NSF allow the number of FTEs allocated to SRS to grow so that the division may broaden the range of staff expertise through hires for new positions as well as through staff turnover.

SRS's budget is substantially smaller than those of other federal statistical agencies and may need to be increased given the growing importance of its subject area and our recommendations for new processes, data collection activities, and additional studies. We did not have access to sufficiently detailed budget data to conduct a cost-benefit analysis either of the items we recommend or of existing components of the SRS portfolio. Thus, we have not been able to prioritize all of our recommendations, nor have we been able to suggest trade-offs between new activities and existing ones. Any budgetary increase must be based on clearer information from SRS on its allocation of internal staff and financial resources across its surveys and other activities, and on a clearer sense of priorities among current and future surveys and activities as developed in coordination with advisory committees for its individual surveys and the SBE Advisory Committee's SRS Breakout Group.

Improving Data Relevance

Science and engineering, a $247 billion enterprise in the United States, plays a central role in the advancement of our knowledge-intensive economy and it affects the daily lives of Americans in myriad ways.[1] The funding, organization, and conduct of science and engineering continue to evolve even as they contribute to economic and social change. To keep its data relevant for answering today's questions on science and engineering resources, SRS needs to keep its data collection portfolio current with these changes. SRS should investigate under-addressed issues in graduate education, the labor market for scientists and engineers, and research and development funding and performance. SRS should use the results of these investigations to revise survey instruments and issues for analysis.

Graduate School and the Transition to Employment

Recommendation 8. SRS should revise its data collection on issues in graduate education and the job market for new Ph.D.s to better address issues on financial support for graduate students, completion of graduate school, and the transition of new Ph.D.s to employment. SRS must carefully study whether fielding a new longitudinal survey of beginning graduate students, now under consideration, is feasible and cost-effective before committing to such a survey. However, the division should revise the Survey of Earned Doctorates (SED) to include questions on progress toward degree completion and job market experiences, and it should seek to assist professional societies and universities in the collection of standardized data on the job market for new Ph.D.s.

Graduate Education

In the face of a difficult job market for recent science Ph.D.s in some fields in the 1990s, policymakers, educators, and analysts have expressed concerns about the efficacy of certain types of support for graduate students and about the outcomes of graduate education. To better understand these issues, they have expressed a desire for additional data on graduate school completion and attrition, career expectations, educational experiences and skills acquired, packages of student financial support, and the effect of each of these on graduate school and career outcomes.

We recommend that SRS analyze and quickly disseminate the retrospective data it collected through the 1997 SDR on the graduate school and job market experience of Ph.D.s who received their degrees between June 1990 and June 1996. These data should address some of the concerns of policymakers, educators, and analysts.

SRS is currently in the development stage for a new longitudinal survey of beginning graduate students, designed to obtain data on education and job market experiences of graduate students. SRS should carefully consider the feasibility and cost-effectiveness of developing and administering such a survey. Based on our current understanding, we question the wisdom of such a survey. SRS should investigate the current state of research on the graduate school experience, attrition and completion, graduate student financial support, and graduate school outcomes. The division should commission additional studies on these subjects as needed to supplement existing research. SRS should then weigh whether the issues warrant ongoing national data collection from graduate students, examine the ability to collect high-quality national data on attrition and packages of financial support, investigate sampling options, determine the cost-effectiveness of conducting a survey that would gather these data, and consider

---

[1] The aggregate amount spent by the federal government, industry, nonprofits, and others on research and development in the United States in 1999.

alternative sources of data. If fielded, the survey would ideally be longitudinal in nature. However, SRS has not fully exploited the longitudinal nature of its other surveys and resources should be committed to a longitudinal survey only if SRS intends to support and utilize it as such.

We do not recommend that such a survey, if fielded, or the Survey of Earned Doctorates be used to collect data on "skills" obtained by graduate students. We do, however, recommend that SRS revise the Survey of Earned Doctorates to obtain data on progress through graduate school, perhaps by adding a question asking respondents for the date when all Ph.D. requirements except for the dissertation were completed.

Transition to Employment

A serious gap in SRS data on science and engineering Ph.D.s has been the job market experience of Ph.D.s in the twelve months before and after receipt of the degree. SRS added retrospective questions on this subject, too, to a one-time module in the 1997 SDR. The division should analyze and disseminate these data, but SRS should also institute ongoing collection of these data by adding questions to the Survey of Earned Doctorates about the job market experience of Ph.D.s prior to degree receipt, and about the salaries of new Ph.D.s who have firm commitments for employment at the time the degree is received.

The division should also continue to augment its own data collection and dissemination by assisting others collecting data in this area—particularly professional societies that survey their members who are recent Ph.D.s and colleges and universities that track the career outcomes of recent science and engineering alumni. The Association of American Universities recently urged research universities to collect data on degree completion by students and on job placement for alumni. If data collected by

these institutions were standardized they could be productively aggregated at the national level.

The Labor Market for Scientists and Engineers

Recommendation 9. SRS should revise its data collection on the labor market for scientists and engineers to better capture the career paths of scientists and engineers. SRS should fill gaps in existing data on careers by collecting comparative data on the careers of humanities doctorates, and data on the nonacademic careers of scientists and engineers, on science and engineering field of work, and on the international mobility of scientists and engineers. The division should also work with the Special Emphasis Panel for the Doctorate Data Project to address content and design issues for the SESTAT system to be implemented in the next decade.

Career Paths of Scientists and Engineers

To improve its data on the labor market for scientists and engineers, SRS should refine or augment several aspects of its personnel surveys to better capture the career paths of scientists and engineers. First, the division should exploit the longitudinal nature of its personnel surveys, which were obtained at great expense and with a respondent burden that is difficult to justify if the data are not used longitudinally. Second, SRS could provide better career path data by making it available at a more detailed level by field. SRS should consider the options available for allowing fine field analysis (for degree field) that is currently obstructed by the small sample size. Increasing sample size is potentially costly, but other options may present themselves. On a related note, science and engineering field of work, dropped from the SDR in 1993 and replaced by a question on occupation, should be restored to the questionnaire. Third, SRS should explore opportunities for linking its personnel data to other career and productivity data, such as data sets of federal research grants, patents, and publications. Fourth, the hole left by the demise of the humanities component of the Survey of Doctorate Recipients seriously undermines our ability to analyze the Ph.D. labor market. SRS must work with the National Endowment for the Humanities (NEH) and other funding sources, if necessary, to reinstate this SDR component.

To better understand the career paths of scientists and engineers and the career options of new Ph.D.s, SRS should revise the SDR to obtain data that better describe the careers of Ph.D.s who work for government agencies, private businesses, and nonprofit organizations. Questions that might be added to better capture nonacademic careers include questions on non-salary compensation; patenting and other productivity measures in the private sector; use of scientific background in sales, regulation, or patent law positions; and temporary work arrangements like contracting and consulting. This is not an exhaustive, but rather an illustrative list.

As the decade comes to a close, we also strongly recommend that SRS work with the Special Emphasis Panel for the Doctorate Data Project to address content and design issues for the SESTAT system for the next decade. The 1989 NRC Report *Surveying the Nation's Scientists and Engineers* recommended that an advisory committee review SESTAT at this time.

International Flows of Scientists and Engineers

Given the globalization of the science and engineering labor market, SRS should develop a long-range plan for improving and increasing the data it collects or acquires on the international flows of scientists and engineers. SRS should begin with an effort to improve data on foreign scientists and engineers at all levels in the United States— students, postdoctorates, and employees. SRS should also examine the costs and benefits of including foreign-educated scientists and engineers working in the United States in the sampling frames for the personnel surveys.

R&D Funding and Performance

Recommendation 10. SRS should revise the data it collects on R&D funding, performance, outputs, and outcomes to improve comparability across surveys and to address structural changes in the science and engineering enterprise. SRS should begin by addressing structural changes in industrial research and development, the relationship between R&D and innovation, the apparent increase in intra- and inter-sectoral partnerships and alliances, and claims that interdisciplinary research is increasing. SRS should examine the costs and benefits of administering the Survey of Industrial Research and Development at the line of business level. SRS should also revise its surveys to address new concepts (e.g., the federal science and technology budget), discrepancies in results among R&D surveys, and the need to obtain better data on academic R&D facility costs.

Industrial R&D Statistics

SRS should improve the accuracy of detailed data on industrial R&D. Currently the Survey of Industrial Research and Development (RD-1), which is fielded at the firm level, attempts to disaggregate both applied research and development by asking respondents to distribute these by product group. Firms, however, often ignore this question and the low response rate to product group has made the collected data of little use. We recommend eliminating the product group question from RD-1. As an alternate strategy for obtaining finer detail on industrial R&D,

SRS should examine the costs and benefits of administering RD-1 to business units instead of firms. Currently all R&D conducted by a firm is attributed to the firm's predominant industrial category. In an economy dominated by large, multi-product firms, line of business reporting, if feasible, may improve data by obtaining finer detail by industrial classification and geographic location.

R&D and Innovation

Current R&D expenditure data do not provide adequate information on many activities contributing to innovation. These activities may include hiring personnel or consultants with new skills, contracting with specialized firms, training existing staff, or reorganizing business processes. SRS should pursue plans to develop a survey of industrial innovation that addresses these and other issues regarding the manner in which science and technology are transferred among firms and transformed into new processes and products. SRS should include both potential respondents and data users in the development of the survey instrument.

SRS should also conduct or sponsor research on the nature of R&D in the service sector. The service sector has increased its share of national R&D investment significantly from less than 5 percent in the early 1980s to almost 25 percent today. SRS should seek to better understand the processes and outcomes of service sector R&D in order to determine the kinds of changes that should be implemented in the Survey of Industrial Research and other surveys to capture important characteristics of R&D in this sector. As a supplement to this investigation, SRS should also examine how personnel data may be used to examine trends in research utilization and innovation.

Partnerships and Alliances

Recent trends suggest that the organizational structure of research and development now includes a web of partnerships and subcontracts among firms, universities, and federal agencies and labs. Yet the extent of such partnerships and their value is not fully understood, in part because of limitations in SRS data. SRS should investigate the nature and variety of these strategic alliances in R&D, including the role of partnerships, outsourcing, mergers and acquisitions, investments in allied firms, and cross-sectoral consortia in performing and supplying R&D. Results of these investigations should guide SRS in revising its survey questionnaires, as necessary, to obtain more complete detail on the role of these partnerships in the science and engineering enterprise. For example, anecdote has suggested that collecting data on inter-sectoral partnerships through the Survey of Industrial R&D may be difficult. A study might explore whether such data could instead be obtained through the Survey of R&D Expenditures in Universities and Colleges and through the Survey of Doctorate Recipients.

Interdisciplinary Research

Similarly, it has become almost a cliché to say that the amount of interdisciplinary and multidisciplinary research is increasing and that much cutting-edge research falls into this category. It is difficult to assess, however, how much research is multidisciplinary because of limitations in SRS data. Given the potential import for such a development on how federal funds should be allocated, the division should hold a workshop or commission a study to better understand the nature and extent of this phenomenon. The workshop or study should be designed to provide insight on how to implement changes consistently across its R&D and human resources data collection efforts in order to better capture multidisciplinary R&D when it occurs.

Allocating Federal Resources for Science and Technology

The allocation of federal resources for science and technology has been much discussed in the wake of the Cold War and substantial increases in funding for the biomedical sciences. At the request of Congress, an NRC panel examined processes for allocating federal resources for research and development and suggested that Congress and the Executive Branch focus on funding trends for the "federal science and technology budget" (FS&T), or that part of R&D spending that focuses on the "creation of new knowledge or technologies" and excludes the testing and evaluation of new weapons systems. SRS could take steps to better support analysis of this concept by requesting that the Department of Energy (DOE) and the National Aeronautics and Space Administration (NASA) break out the FS&T portion from their aggregate budget and obligation figures as does the Department of Defense (DOD).

SRS should also continue to takes steps to investigate and reconcile discrepancies in R&D funding data obtained by its different surveys since they hamper analyses of federal funding. For example, the Survey of Federal Funds for Research and Development estimates that federal R&D obligations to academic institutions in 1997 was $12.2 billion while the Survey of R&D Expenditures at Universities and Colleges estimates federally-funded R&D expenditures in that year to be $14.1 billion. Similarly, the Federal Funds Survey estimates a decrease of 32 percent in federal obligations to academic institutions for electrical engineering research between 1993 and 1997, while the academic R&D survey estimates for that period an increase in federally-funded academic R&D in that field of 27 percent. Some of this discrepancy may be accounted for in the difference between counting research as opposed to research *and* development, but not all. SRS needs to resolve these discrepancies to improve the credibility of the data among analysts.

Academic Facilities

SRS should continue to pursue changes to its Survey of Scientific and Engineering R&D Facilities at Colleges and Universities to provide better data for assessing overhead rates at research universities and estimating future academic infrastructure needs. The collection of improved data on academic facilities to assist OMB in this effort is an important test of SRS's ability to provide data relevant to policymaking.

References

American Association for the Advancement of Science (AAAS)
1999   *Research and Development FY 2000: AAAS Report XXIV.* Washington, D.C.: American
        Association for the Advancement of Science.

Association of American Universities (AAU)
1995   *AGS Statement on the COSEPUP Report: Reshaping the Graduate Education of
        Scientists and Engineers.* Washington, D.C.: Association of American Universities.

1998   *Committee on Graduate Education: Report and Recommendations, October 1998.*
        Washington, D.C.: Association of American Universities.

Atkinson, Robert D. and Randolph H. Court
1998   *The New Economy Index: Understanding America's Economic Transformation.*
        Washington, D.C.: Progressive Policy Institute.

Bloom, Floyd E.
1998   "Priority Setting: Quixotic or Essential?" *Science*, November 27, 1998, 1641.

Boesman, William C.
1997   "Analysis of Ten Selected Science and Technology Policy Studies," 97-836.
        Washington, D.C.: Congressional Research Service.

Bowen, William G., and Neil L. Rudenstine
1992   *In Pursuit of the Ph.D.* Princeton, NJ: Princeton University Press.

Bush, Vannevar
1945   *Science: The Endless Frontier: A Report to the President.* Washington, D.C.: United
        States Government Printing Office.

Champion, Elinor J.
1998   "The Survey of Industrial Research and Development: Overview and Recent Developments." Presentation to the Census Advisory Committee of Professional Associations. April 23, 1998.

Committee for Economic Development
1998   *Basic Research: Prosperity Through Discovery.* New York, NY: Committee for Economic Development.

Cox, Brenda G., Susan B. Mitchell, and Ramal Moonesinghe
1998a   *Current and Alternative Designs for the Survey of Doctorate Recipients.* Washington, D.C.: Mathematica Policy Research, Inc.

1998b   *Using the Survey of Doctorate Recipients in Time-Series Analyses: 1989-1995.* Washington, D.C.: Mathematica Policy Research, Inc.

Council on Competitiveness
1998   *Endless Frontier, Limited Resources: U.S. R&D Policy for Competitiveness.* Washington, D.C.: Council on Competitiveness.

Davis, Paul W., James W. Maxwell, and Kinda Remick
1998   "1997 Annual Survey of the Mathematical Sciences: First Report." *Notices of the AMS.* February 1998.

1999   "1998 Annual Survey of the Mathematical Sciences: First Report." *Notices of the AMS.* February 1999.

Hagedoorn, John
1996   "The Economics of Cooperation Among High-Tech Firms: Trends and Patterns in Strategic Technology Partnering Since the Early Seventies." In George Koopmenn and Hans-Eckart Scharer, ed. *The Economics of High-Technology Competition and Cooperation in Global Markets.* Baden-Baden: Nomos Verlagsgesellschaft.

Hansen, John A., James I. Stein, and Thomas More (with contributions from Christopher Hill and James H. Maxwell)
1984   *Industrial Innovation in the United States—A Survey of 600 Companies* (NSF 84-1)

Hill, Christopher T., John A. Hansen, and James I. Stein
1983   *New Indicators of Industrial Innovation* (CPA-83-14)

Metzger, Norman and Richard Zare
1999   "Interdisciplinary Research: From Belief to Reality." *Science* (283):642-643.

Mowery, David C.
1999   "America's Industrial Resurgence: How Strong, How Durable?" *Issues in Science and Technology Policy,* Spring 1999, 41-48.

112

Mulvey, Patrick J.
1998   "1997 Initial Employment Report: Follow-up of 1996 Physics Degree Recipients," *AIP Report,* July 1998.

National Academy of Sciences, National Academy of Engineering, and Institute of Medicine, Committee on Science, Engineering, and Public Policy (NAS)
1995   *Reshaping the Graduate Education of Scientists and Engineers.* Washington, D.C.: National Academy Press.

1999   *Observations on the President's Fiscal Year 2000 Federal Science and Technology Budget.* Washington, D.C.: National Academy Press.

National Research Council
1951   "The Doctorate Survey." Washington, D.C.: Office of Scientific Personnel.

1989   *Surveying the Nation's Scientists and Engineers: A Data System for the 1990s.* Washington, D.C.: National Academy Press.

1992   *Principles and Practices for a Federal Statistical Agency.* Washington, D.C.: National Academy Press.

1993   *Validation Study of the Survey of Earned Doctorates.* Washington, D.C.: Office of Scientific and Engineering Personnel.

1994   *Allocating Federal Funds for Science and Technology.* Washington, D.C.: National Academy Press.

1996a   *The Path to the Ph.D.: Measuring Graduate Attrition in the Sciences and Humanities.* Washington, D.C.: National Academy Press.

1996b   *Summary Report 1995: Doctorate Recipients from United States Universities.* Washington, D.C.: National Academy Press.

1997a   *Bureau of Transportation Statistics: Priorities for the Future.* Washington, D.C.: National Academy Press.

1997b   *Humanities Doctorates in the United States: 1995 Profile.* Washington, D.C.: National Academy Press.

1997c   *Industrial Research and Innovation Indicators: Report of a Workshop.* Washington, D.C.: National Academy Press.

1998a   *Doctoral Scientists and Engineers in the United States: 1995 Profile.* Washington, D.C. National Academy Press.

1998b   *Summary Report 1996: Doctorate Recipients from United States Universities.* Washington, D.C.: National Academy Press.

1998c *Trends in the Early Careers of Life Scientists.* Washington, D.C.: National Academy Press.

1999 *Securing America's Industrial Strength.* Washington, D.C.: National Academy Press.

National Science Board
1995 *Science & Engineering Indicators—1996.* (NSB 96-21) Arlington, VA: National Science Foundation.

1997 *Government Funding of Scientific Research,* (NSB-97-186) Arlington, VA: National Science Foundation.

1998a *Science & Engineering Indicators—1998.* (NSB 98-1) Arlington, VA: National Science Foundation.

1998b *Strategic Plan.* (NSB-98-215). Arlington, VA: National Science Foundation.

1998c *The Federal Role in Science and Engineering Graduate and Postdoctoral Education.* (NSB 97-235) Arlington, VA: National Science Foundation.

National Science Foundation
1991 *Characteristics of Doctoral Scientists and Engineers in the United States: 1989: Detailed Statistical Tables.* (NSF 91-317) Washington, D.C.: National Science Foundation.

1994 "Customer Views of SRS Products and Services." Customer Service Task Force, Division of Science Resources Studies, National Science Foundation. June 8, 1994.

1996 "Data for Monitoring and Analyzing Graduate Education: A Report to the Data Needs Committee of the SMIG." Internal Memorandum by the Data Group on Graduate Education. July 1996.

1997 Memorandum, January 23, 1997. From Alan R. Tupek, Deputy Director, Division to Science Resources Studies, to All SRS Staff. "Results of SRS Customer Survey."

1998a *Division of Science Resources Studies, Fiscal Year 1999 Budget Call.* Arlington, VA: National Science Foundation.

1998b "How Has the Field Mix of Academic R&D Changed?" *Division of Science Resources Studies Issue Brief.* (NSF 99-309) December 2, 1998.

1998c "International Mobility of Scientists and Engineers to the United States—Brain Drain or Circulation?" *Division of Science Resources Studies Issue Brief.* (NSF 99-316) June 22, 1998; revised November 10, 1998.

1998d "What are the Sources of Funding for Academically Performed R&D?" *Division of Science Resources Studies Issue Brief.* (NSF 99-317) December 23, 1998.

1999a *Academic Research and Development Expenditures, Fiscal Year 1997.* (NSF 99-336). Arlington, VA: National Science Foundation.

1999b *Federal Funds for Research and Development: Fiscal Years 1997, 1998, and 1999.* (NSF 99-333). Arlington, VA: National Science Foundation.

1999c *Federal Funds Survey, Detailed Historical Tables: Fiscal Years 1951-99.* (NSF 99-347). Arlington, VA: National Science Foundation.

1999d *Federal R&D Funding by Budget Function: Fiscal Years 1997-99.* (NSF 99-315). Arlington, VA: National Science Foundation.

1999e *Graduate Students and Postdoctorates in Science and Engineering: Fall 1997.* (NSF 99-325) Arlington, VA: National Science Foundation.

1999f *National Patterns of R&D Resources: 1998.* (NSF 99-325) Arlington, VA: National Science Foundation.

1999g *Research and Development in Industry: 1995-1996.* (NSF 99-312). Arlington, VA: National Science Foundation.

1999h *Research and Development in Industry: 1997 [Early Release Tables].* Available at http://www.nsf.gov/sbe/srs/srs99411/start.htm (will be published later in 1999).

1999i *Science and Engineering Doctorate Awards: 1997.* (NSF 99-323). Arlington, VA: National Science Foundation.

1999j *Scientific and Engineering Research Facilities at Colleges and Universities: 1998: An Overview.* (NSF 99-413) Arlington, VA: National Science Foundation.

1999k *SESTAT: A Tool for Studying Scientists and Engineers in the United States* (NSF 99-337). Arlington, VA: National Science Foundation.

1999l "What is the Federal Role in Supporting Academic Research and Graduate Research Assistants?" *Division of Science Resources Studies Issue Brief.* (NSF 99-342) April 16, 1999.

Spar, Edward J.
1999 "Federal Statistics in the FY 2000 Budget," in Intersociety Working Group, *Research and Development FY 2000: AAAS Report XXIV.* Washington, D.C.: American Association for the Advancement of Science.

Stephan, Paula E.
1996 "The Economics of Science." *Journal of Economic Literature* (34):1199-1235.

U.S. Congress. House. Committee on Science. *Unlocking our Future: Toward a New*
1998 *National Science Policy.* September 24, 1998.

U.S. Office of Management and Budget (OMB)
1999 *The President's FY 2000 Budget.* Washington, D.C.: U.S. Government Printing Office.

Appendix A

Interview and Focus Group Guides

Surveying Resources for Science and Engineering:
Assessing the Sciences Resources Studies (SRS) Portfolio

Part A:

Questions for SRS Survey Project Officers

General Information about SRS Staff:

Name: _____

Primary Survey: **(A)**_____

Other Surveys: **(B)**_____

                 (C)_____

How long have you been with SRS? _____ Years

How long have you worked on these surveys? (A:)_____ Years
                                  (B:)_____ Years
                                  (C:)_____ Years

Attach a brief description of what the interviewee does (analysis, overseeing the work of the survey contractor, etc.)—obtained from the interviewee before the interview.

INTRODUCTION. (Read to the interviewee:) *We're interested in asking you about the SRS surveys you work with and how well you think they address important policy issues and research questions involving science resources. The topics we want to cover are summarized below. Please be assured that your replies will be confidential—we will not make available outside our study any information that would link your name with your responses.*

Question Area 1. SURVEY CONTENT. (Read to the Interviewee:) *We're interested in your assessment of the relevance of the survey content to important policy and research issues.*

(Ask for each survey:)

➤ Which policy and research issues do the questions on your survey(s) address and which issues are most important in your view?

➤ Are different/additional survey questions needed to address these issues more fully?

➤ What other current or emerging policy and research issues might your survey address that it does not address now?

➤ Could the survey add useful questions on these topics?

➤ What would be the survey questions/topics you would suggest adding, in priority order?

➤ Would these questions/topics need to be added permanently or just one-time?

➤ Conversely, are there questions that the survey could drop because they're obsolete?

Question Area 2. SURVEY QUALITY. (Read to the interviewee:) *We're interested in your assessment of the quality of the survey along several dimensions and the possible effects of quality concerns on the policy and research uses of the data. Areas of quality concern include the definition of the target population, coverage of the population, sample design, nonresponse, response errors, frequency of data collection, and timeliness of data products that may affect the use of the data for policy or research.*

(Ask for each survey:)

➤ What quality concerns or "problems" affect the surveys you work with and the data they obtain?

➤ How severe are these problems in your view?

➤ What methodological or operational improvements would you suggest implementing for the survey that would improve its usefulness, in priority order?

Question Area 3. FIT WITH OTHER SURVEYS. (Read to the interviewee:) *We're interested in your assessment of how well the survey fits with other surveys carried out by SRS or other agencies.*

(Ask for each survey:)

➢　　In what ways could coordination of the survey with other surveys on science and engineering resources be improved?

Question Area 4. DATA USERS AND DATA USES. (Read to the interviewee:) *We're interested in your assessment of the audience for the survey and the publications and other products based upon them.*

(Ask for each survey:)

➢　　Who uses the survey? Are there different types of users for the survey?

➢　　How well does the survey serve the needs of these different types of users in terms of content, accuracy of data, level of detail, timeliness, and forms of dissemination?

➢　　What improvements would you suggest implementing, in priority order, that would make the data more useful to current users?

　　➢　　....to users not now being reached?

Question Area 5. USE OF DATA IN PUBLICATIONS AND ANALYSES. (Read to the interviewee:) *We're interested in your assessment of how well the data from the survey are used in SRS publications, including Science and Engineering Indicators.*

(Ask for each survey:)

➢　　Are the data used descriptively?

　　➢　　....analytically?

➢　　Are the data analyzed appropriately?

➢　　What additional analyses would you suggest, in priority order, that SRS should conduct and publish of the data?

Question Area 6. OVERALL ASSESSMENT OF SRS'S PROGRAM. *Finally, we're interested in other comments you may have on SRS's program and how well it meets users' needs for data for policy analysis and research.*

Surveying Resources for Science and Engineering:
Assessing the Sciences Resources Studies (SRS) Portfolio

Part B:

Interview Guide for Science and Technology Policy Analysts and Researchers
(includes data analysts who are SRS staff)

General Information about Data Users

Name: _____

Organization: _____

Primary Survey Used: (A)_____

Other Surveys Used: (B)_____

(C)_____

How long have you used SRS Surveys? _____ Years

How long have you used specific surveys? (A:)_____ Years
(B:)_____ Years
(C:)_____ Years

Attach a brief description of the interviewee's role, how s/he uses science and technology resources data, and the data sources s/he uses—obtained from the interviewee before the interview.

INTRODUCTION. (Read to the interviewee:) *We're interested in asking you about the kinds of policy analysis or research you conduct on science and technology resources issues, the data you use, and their adequacy for your analysis purposes. We're particularly interested in your views about the usefulness of data from the National Science Foundation's Science Resources Studies Division (which we will refer to as SRS), but if you use data from other sources, we would like to hear about those data, too.*

Please be assured that your replies will be confidential—we will not make available outside our study any information that would link your name with your responses.

Question Area 1. POLICY/RESEARCH ISSUES AND DATA SOURCES. (Read to the Interviewee:) *We're interested in knowing more about the policy or research issues you work with and the data sources you use.*

(Ask:)

➢ Please tell us the key policy and/or research issues that you are interested in and work on. (If obtained prior to the interview, ask the interviewee to elaborate as necessary and possible).

➢ What kinds of data do you use to work on these issues and what are their sources? (If a list of data sources was obtained prior to the interview, ask the interviewee to elaborate as necessary and possible).

➢ From among the data sources you tap, which SRS surveys—and which summaries and tables within publications using SRS data—address policy or research issues that are important to your work and which issues?

➢ If you don't use SRS data, why not? (Probe to see if s/he uses SRS data but doesn't know it.)

Question Area 2. SURVEY CONTENT. (Read to the Interviewee:) *We're interested in your assessment of how relevant existing data on science and technology resources—from SRS and other sources—is to your analysis and research.*

(Ask:)

➢ What is your assessment of the overall content of the surveys that you use and its relevance to your policy or research issues?

➢ Do these surveys present data at the level of detail you require?

➢ What current or emerging policy and research issues you are concerned about are not addressed by the SRS and other surveys that you use?

➢ Could useful survey questions on these topics be added to these surveys or a new survey? If so, what would be the survey questions/topics you would suggest adding to which surveys, in priority order?

➢ Would these questions/topics need to be added permanently or just one-time?

➢ Are there questions that any of these surveys could drop because they're obsolete?

Question Area 3. SURVEY QUALITY. (Read to the interviewee:) *We're interested in your assessment of the quality of the survey data you use along several dimensions and the possible effects of quality problems on the policy and research uses of the data. Areas of quality concern that may affect the use of the data for policy or research include the definition of the target population, coverage of the population, sample design, nonresponse, response errors, frequency of data collection, or the timeliness of data products.*

(Ask:)
➢ What quality problems affect the survey data you work?

➢ How severe are these problems in your view?

➢ What methodological improvements would you suggest implementing for the surveys that you use that would improve their usefulness—in priority order?

Question Area 4. FIT WITH OTHER SURVEYS. (Read to the interviewee:) *We're interested in your assessment of how well data from the SRS surveys that you use fit with other surveys conducted by SRS or other agencies and organizations.*

(Ask:)
➢ In what ways could coordination of SRS surveys with other surveys be improved to provide data that can be integrated in a way that would be useful in your work?

➢ Do you see any unnecessary overlaps or redundancies in SRS surveys and other surveys you use?

➢ Are there any specific inconsistencies between SRS and other surveys in the ways that data are collected that make the data from these surveys difficult to integrate and use fully?

Question Area 5. ACCESS TO DATA AND PUBLICATIONS. (Read to the interviewee:) *We're interested in how well you are able to access the data and publications you need for your work and your assessment of the publications you use.*

(Ask for each survey used by the interviewee:)
➢ How do you typically access data from SRS and other sources? (via the web, data files, customized tables, off-the-shelf publications)

➢ Are you able to access data from these sources in an effective way?

➢ Do you find descriptive and analytical presentation of data in SRS and other publications useful and appropriate?

➢ Would different/additional summaries, tables, or analyses be useful in addressing the policy or research issues you work on? Please specify, in priority order, what SRS and other sources might add.

Question Area 6. FUTURE DATA NEEDS. (Read to the interviewee:) *We're interested in the science and technology issues you see on the horizon that may require different or additional data.*

(Ask for each survey used by the interviewee:)

➢ What new science resources issues do you see on the horizon that may require different or additional data?

➢ What types of data do you anticipate will be required?

➢ Do you anticipate that existing surveys can address these data needs or will these new issues require additional surveys to address them?

➢ What immediate or long-term changes would you propose, in priority order?

Question Area 7. OVERALL ASSESSMENT OF SRS'S SURVEY AND PUBLICATIONS PROGRAM. *We're interested in other comments you may have on SRS's program and how well it meets your needs for data for policy analysis and research.*

Surveying Resources for Science and Engineering:
Assessing the Sciences Resources Studies (SRS) Portfolio

Part C:

Questions for
SRS Information Services Group

INTRODUCTION. (Read to the interviewee:) *We're interested in asking you about SRS data and their collection, analysis, and dissemination. In particular, we want to know how well you think the data, analyses, and publications serve the public, policymakers, and researchers who are interested in science resources.*

Please be assured that your replies will be confidential—we will not make available outside our study any information that would link your name with your responses.

Question Area 1. DATA USERS AND DATA USES. (Read to the interviewee:) *We're interested in your assessment of the audience for SRS data and other data on science and engineering resources:*

➢ Who uses SRS data? What different types of users can you describe?

➢ Who doesn't use SRS data that, in your opinion, should be reached? Why aren't they reached—why don't they use SRS data?

➢ How well does SRS data serve the needs of these different types of users in terms of content, accuracy of data, level of detail, timeliness, and forms of dissemination?

➢ How do they obtain SRS data—publications, tables, special requests, web usage, etc.? How is access to data an issue?

➢ What improvements would you suggest implementing that would make the data more useful to current users?

    ➢ Different data? Different analyses? Different means of accessing data?

Question Area 2. USE OF DATA IN PUBLICATIONS AND ANALYSES. (Read to the interviewee:) *We're interested in your assessment of how well the data from the survey are used in SRS publications, including Science and Engineering Indicators.*

➢ For those users who do access SRS data in one form or another, are they finding what they want? (i.e., raw data, data tables, brief analyses, charts and graphs, analytical reports, etc.?

➢ What different or additional uses of the data—or certain data sets—would you suggest, in priority order, that SRS consider?

Question Area 3. SURVEY CONTENT. (Read to the Interviewee:) *We're interested in your assessment of the relevance of the survey content to important policy and research issues.*

➢ To what extent can you identify areas in which the relevance or currency of certain SRS data are issues in their use? Which policy and research issues do users find information on? Which issues do they not find information on?

➢ Are different/additional survey questions needed to address these issues more fully?

➢ What other current or emerging policy and research issues on science and engineering resources might SRS other data sources that it does not address now?

➢ Could useful survey questions be added on these topics? What would be the survey questions/topics you would suggest adding, in priority order? Would these questions/topics need to be added permanently or just one-time?

➢ Conversely, are there questions that the survey could drop because they're obsolete?

Question Area 4. SURVEY QUALITY. (Read to the interviewee:) *We're interested in your assessment of the quality of the survey along several dimensions and the possible effects of quality concerns on the policy and research uses of the data. Areas of quality concern include the definition of the target population, coverage of the population, sample design, nonresponse, response errors, frequency of data collection, and timeliness of data products that may affect the use of the data for policy or research.*

➢ What quality concerns or "problems" affect SRS data you work with?

➢ How severe are these problems in your view?

➢ What methodological or operational improvements would you suggest implementing for the survey that would improve its usefulness, in priority order?

Question Area 5. FIT WITH OTHER SURVEYS. (Read to the interviewee:) *We're interested in your assessment of how well SRS surveys fit with each other and with surveys carried out by other agencies.*

➢ In what ways could coordination of SRS surveys with each other and other surveys on science and engineering resources be improved?

Question Area 6. OVERALL ASSESSMENT OF SRS'S PROGRAM. *Finally, we're interested in other comments you may have on SRS's program and how well it meets users' needs for data for policy analysis and research.*

Appendix B

Workshop Agenda

AGENDA

Workshop on Data to Describe Resources for
the Changing Science and Engineering Enterprise

September 18-19, 1998

Conference Room 130
Cecil and Ida Green Building
2001 Wisconsin Avenue, N.W.
Washington, DC 20007

NATIONAL RESEARCH COUNCIL
Office of Scientific and Engineering Personnel
Committee on National Statistics

Friday, September 18, 1998

8:00 a.m.   **Continental Breakfast**

8:30 a.m.   **Welcome**

Janice Madden, University of Pennsylvania
Jeanne Griffith, National Science Foundation
Charlotte Kuh, National Research Council

8:45 a.m.   **Federal Policy and the Changing Science and Engineering Enterprise**

Moderator:
Robert McGuckin, The Conference Board

Speaker:
J. Thomas Ratchford, Center for Science, Trade, and Technology Policy,
     George Mason University

9:15 a.m.   **Issues and Data for Science and Technology Policy**

Moderator:
Paul Biemer, Research Triangle Institute

Speakers:
Eric Fischer, Congressional Research Service
Patrick Windham, R. Wayne Sayer and Associates
Ray Merenstein, Research!America

10:30 a.m.   **Break**

10:45 a.m.   **Measuring the Global Dimensions of Science, R&D and Innovation**

Moderator:
David Mowery, University of California, Berkeley

Speakers:
Gerald Hane, White House Office of Science and Technology Policy
Daniel Malkin, Organization for Economic Cooperation and Development
Fred Gault, Statistics Canada

12:00 noon   Lunch *(box lunches and beverages available in Room 128)*

12:30 p.m.   "The Role of a Federal Statistical Agency"

Luncheon speaker:
Katherine Wallman, Chief Statistician
U.S. Office of Management and Budget

1:00 p.m.   Measuring Education, Training, and Careers in Science and Engineering

Moderator:
Paula Stephan, Georgia State University

Speakers:
Denice Denton, University of Washington
Leslie B. Sims, University of Iowa
Joseph Jasinski, IBM Research
Susanne Huttner, University of California System

2:30 p.m.   Research Universities and the Changing Science and Engineering Enterprise

Moderator:
Julie Norris, Massachusetts Institute of Technology

Speakers:
Charles Vest, Massachusetts Institute of Technology
Linda Cohen, University of California, Irvine

3:30 p.m.   Break

3:45 p.m.   Measuring International Flows of Scientists and Engineers

Moderator:
T.R. Lakshmanan, Boston University

Speakers:
Michael Teitelbaum, Sloan Foundation
Michael Finn, Oak Ridge Institute for Science and Education
Dominique Martin-Rovet, Science and Technology Mission, Embassy of France

5:00 p.m.   Reception

Saturday, September 19, 1998

8:00 a.m.   **Continental Breakfast**

8:30 a.m.   **The Changing Science and Engineering Enterprise in Industry**

Moderator:
John McTague, Ford Motor Company

Speaker:
Stephen J. Lukasik, Independent Consultant
Edward Penhoet, University of California, Berkeley

9:30 a.m.   **Measuring the Changing Science and Engineering Enterprise in Industry**

Moderator:
Bronwyn Hall, University of California, Berkeley

Speakers:
Irwin Feller, Pennsylvania State University
John Birge, University of Michigan

10:30 a.m.   **Break**

10:45 a.m.   **New Directions in Federal Science and Technology Policy**

Moderator:
Eduardo Macagno, Columbia University

Speakers:
Arthur Bienenstock, White House Office of Science and Technology Policy
Shirley Malcom, American Association for the Advancement of Science

11:45 a.m   **Adjourn**

Appendix C

Surveying the Nation's Scientists and Engineers
Recommendations

Surveying the Nation's Scientists and Engineers:
A Data System for the 1990s **(NRC 1989)**

Recommendations

Chapter 5: Priority Goals and Design Features of the System for the 1990s

5.1 NSF should continue to be the lead agency within the federal government for providing comprehensive data on the science and engineering personnel resources of the nation. NSF must undertake to provide the budget and staff resources and institutional support necessary to develop and maintain a personnel data system that will adequately meet the needs of the 1990s and beyond.

5.2 Other federal agencies will continue to collect data in support of their own missions that pertain to science and engineering personnel. In order to enhance data comparability and utility to the extent practicable and to reduce duplication of effort and costs, NSF should play the lead role in coordinating federal data programs on scientists and engineers. Within the framework of established federal classification schemes, NSF should encourage standardization of key questionnaire items and classification variables for science and engineering personnel across agencies.

5.3 Currently, the primary goal of the NSF data system is to provide information on the characteristics of science and engineering personnel in order to support the planning processes of government, academic, and business institutions. In the 1990s, the data system should continue to serve this goal. Specifically, the system should

- Support the preparation of regular profiles of the characteristics of scientists and engineers, including their numbers, employment patterns, qualifications, utilization, and other characteristics, with separate tabulations by field, sex, and race, when feasible, and
- Support the preparation of special analyses that illuminate specific policy issues and characteristics of science and engineering personnel in greater depth

In the 1990s, the data system should also serve other important goals to which NSF does not currently accord high priority:

- Provide a research base for improved analysis of relevant labor markets and of flows into, out of, and within the science and engineering labor force that can pinpoint trouble spots and provide early warnings of future problems, and
- Provide a database that will support basic innovative research on scientists and engineers and the science and engineering pipeline.

5.4 NSF should provide information about the full range or people who can be considered as part of the science and engineering supply. NSF should furnish information on the population of graduates in science and engineering fields, not all of whom have related work experience. NSF should also furnish information on the population of employed scientists and engineers, not all of whom were trained in science and engineering fields. NSF should discard the current screening algorithm as a means of defining the population. Instead, NSF should use definitions based on standard occupation and degree field categories, developing within these frameworks more richly detailed classifications of subgroups of scientists and engineers.

5.5 NSF should increase the research utility of the science and engineering personnel database by enriching the content of its surveys. NSF should assign priority to new or modified content items that will provide greater understanding of:

- The kinds of work that scientists and engineers do and how their work is changing in response to changes in technology, organizational structure, and other factors;
- The career paths that scientists and engineers follow and the factors that influence key transitions, including initial entry into the labor force, mobility across fields and sectors, and retirement; and
- The productivity that scientists and engineers achieve and how their accomplishments relate to characteristics of their training, career moves, and work environment.

5.6 NSF should conduct a large Postcensal Survey of College Graduates in 1992 based on the 1990 decennial census that provides baseline information on college graduates, including those who are trained in science and engineering fields and those with employment in science and engineering occupations.

5.7 NSF should conduct a Panel Survey of Scientists and Engineers that periodically provides updated information on the population of college graduates with science and engineering degrees and that tracks the 1992 cohort of graduates with employment in science and engineering occupations, including those who were trained in other fields. The survey should also include a sample for each new graduating class that is drawn from the Prospective Graduates Surveys conducted each year of students at higher education institutions who are about to receive a bachelor's or master's degree in a science or engineering field.

5.8 NSF should continue to support the ongoing Survey of Doctorate Recipients and employ it as the major source of information on science and engineering personnel trained to the Ph.D. level. The SDR should be modified to facilitate its use with the other surveys in the NSF science and engineering personnel data system.

5.9 To the extent possible during the decade, NSF should use other federal data sources to obtain information on components of the science and engineering population that are not covered in the NSF survey system and to evaluate NSF's survey-based estimates.

5.10 Because of the importance of degree field in defining the population of scientists and engineers, the Current Population Survey should periodically include a supplement that asks respondents for major field of bachelor's and higher degrees. NSF should work for the adoption of this recommendation by the Bureau of Labor Statistics and the Census Bureau.

5.11 NSF should pursue its planned research program to develop estimates of immigration and emigration of scientists and engineers and to develop ways of incorporating such estimates into the personnel data system.

5.12 NSF should consider the study panel's recommended design for its science and engineering personnel data system in the 1990s as a package in which basic information on the population of scientists and engineers, detailed information on topics and subgroups of key analytic interest, and evaluation and augmentation of NSF's own survey estimates using other federal data sources are integral and equally important elements.

Chapter 6: Designing and Implementing the System for the 1990s

6.1 In order to conserve resources and reduce burden on higher education institutions, NSF and the National Center for Education Statistics should design a unified sampling frame and coordinate procedure for obtaining data on prospective college graduates. The two agencies should not combine their panel surveys of new graduates, however, which serve different purposes and focus on different fields of training. Key questionnaire items should be comparable in order to permit each agency to evaluate and supplement its own data with the data from the other agency.

6.2 NSF should initiate modifications to the Survey of Doctorate Recipients, specifically, in the areas of coverage, survey scheduling, sample design, and wording of key questionnaire items, that will improve comparability of the SDR data with other data in the NSF science and personnel data system.

6.3 Detailed specification of the design for the NSF personnel data system in the 1990s will require additional analysis and decision making. NSF should in the near term set a process for reaching final design decisions. This process should include:

- Identifying and funding priority research and analysis projects whose results are needed to inform the design;
- Establishing a group of technical experts to work with NSF staff in reaching final design decisions and to assist NSF in monitoring the operation of the system in the 1990s; and
- Sponsoring workshops and in other ways seeking both to obtain input from users and to advise them of impending changes in the data system.

6.4 Toward the end of the 1990s, NSF should conduct a thorough, zero-based evaluation of the design and operation of its personnel data system to determine whether to continue the basic design of the 1990s or to change the system in important ways, The evaluation should include a review of the goals of the system and the extent to which the informational content is serving those goals.

Chapter 7: Operating the System in the 1990s

7.1 NSF should develop a quality profile for its personnel surveys that will guide the development of an effective system to monitor and maintain data quality and suggest research to learn more about sources of error in the data and to identify further possible improvements.

7.2 NSF should take advantage of the experience of other federal statistical agencies in developing quality profiles, setting quality standards, and implementing quality control programs. NSF should keep abreast of procedures and techniques that federal agencies and private survey research centers use for improving data quality, particularly of data from continuing panel surveys.

7.3 NSF should devote a significant portion of its budget each year for the personnel data system to quality review and improvement activities.

7.4 When faced with budget constraints that necessitate trade-offs, NSF should choose options for the system that minimize total error in the data, taking into account both sampling error and nonsampling error from sources such as nonresponse.

7.5 NSF should adopt the best survey practice in designing and evaluating questionnaires for its science and engineering personnel surveys.

7.6 NSF should adopt the best survey practice for its personnel surveys in the following operational areas:

- Procedures for obtaining high levels of response, both through initial contact and follow-up;
- Procedures for data preparation, including developing appropriate weights, imputing missing values, and editing the data for consistency.

7.7 NSF should provide resources to the Science Resources Studies Division for staff training in survey methodology and for staff to attend conferences, short courses, and other venues of continuing education. NSF should also provide resources for the staff to develop first-hand knowledge, through field visits and other means, of the many different kinds of scientists and engineers whose characteristics the personnel surveys are intended to measure.

7.8 NSF should provide the resources for the staff of the Science Resources Studies Division to have access to the personnel microdata. NSF should encourage the staff to use the data for analytical studies, particularly those that relate to data quality and methodology, and to present their findings at professional meetings and in professional journals.

7.9 NSF should include resources in its survey contracts for contractors to propose and carry through research related to understanding and improving data quality.

Chapter 8: Building a User Community

8.1 NSF should plan an extensive publication program from the 1992 Postcensal Survey, which will provide the first comprehensive look in a decade at the entire population of scientists and engineers and permit comparative analysis with other subgroups of college graduates.

8.2 As a major publication series form the continuing Panel Survey, NSF should regularly publish profiles of college graduates with science and engineering degrees that separately identify important subgroups to permit users to apply a narrow or broad definition of the population as suits their needs. Two basic tabulation series would be useful: one series that focuses on the current employment situation of people with degrees in particular science and engineering fields, and another that focuses on the educational background and work environments of science and engineering graduates who are employed in particular science and engineering occupations. NSF should also produce publications from the Panel Survey about the cohort of employed scientists and engineers (including people trained in other fields) identified in the 1992 Postcensal Survey and about new graduates in science and engineering fields.

8.3 In determining the categorization of degree field, occupation, and other variables in NSF tabulations, user needs for more information must be balanced against considerations of sampling error. NSF should set standards for the minimum size science and engineering field for which estimates will be published based on the error properties of its surveys. Conversely, NSF should seek meaningful ways to provide additional detail for larger science and engineering fields.

8.4 NSF should provide a variety of products from the personnel data system—printed reports, public use microdata files, and other computer-readable products—that serve the needs of the entire user community, ranging from those users who require a few specific numbers to those users who are engaged in extensive analysis.

8.5 NSF should implement the recommendations that are developed by the Committee on National Statistics from its recent effort to seek ways to improve research access to the Survey of Doctorate Recipients while protecting the confidentiality of individual replies.

8.6 NSF should provide complete documentation for all products made available from the personnel data system, including a comprehensive user's guide to accompany public use microdata files. Data file documentation and technical notes included in publications should emphasize the nature and likely magnitude of the errors in the data.

8.7 NSF should actively publicize the availability of public use microdata files and other products from its personnel surveys.

8.8 NSF should actively encourage and provide support to researchers for innovative studies of science and engineering personnel using survey microdata. NSF should consider for this purpose establishing a grants program to fund projects that use the personnel data.

8.9 NSF should actively solicit feedback from its users on the design, content, and quality of the data system, and on the content and format of data products. NSF should consider for this purpose establishing a user panel to provide input on a regular basis.

Appendix D

Alternative Sampling Frames
For SRS Personnel Surveys

Alternative Sampling Frames For Personnel Surveys

Users of data on scientists and engineers commonly complain that it is not possible to identify subgroups in this population in sufficient detail, that there is not enough information about the working environment, career paths, and other characteristics of scientists and engineers, and that trends over time, such as changes in the numbers of science and engineering immigrants, are not well monitored. Statistical problems may, in part, explain insufficient data in these areas. SRS should investigate whether alternative sampling frames for its personnel surveys would make it possible to improve these data in the future.

The Problem

Scientists and engineers are what are known in the survey methods field as a "rare population," which makes them difficult to study in a cost-effective manner. At first glance, this classification appears to be hyperbole. For 1993, the NSF SESTAT system estimated that there were 12 million college graduates under age 76 with either employment or training (or both) in a science and engineering field—not a small group in total numbers. The Current Population Survey (CPS), which is the largest continuing U.S. household survey, has about 48,000 households and 126,000 people in the sample each month, of which an estimated 6,000 people would be scientists and engineers. This sample size is adequate for reliable estimates of the total population of scientists and engineers. (That is, the sample size would be adequate if the CPS permitted identifying people trained but not currently working as scientists and engineers in addition to those employed in science and engineering.)[1]

However, users of science and engineering personnel data are almost never interested in the total: their interests center on particular groups (e.g., women and minorities in specific fields) and comparing across groups. For a group amounting to 50,000 people, which is not atypical for specific fields, the sample size in one month of the CPS is only about 25 cases, which is not adequate for analysis purposes. Also, the total population of working scientists and engineers, while not that small numerically, is a small percentage of the total household population (5% in 1993).

To obtain adequate sample size for a rare population in a cost-effective manner, it is necessary to find some way to screen the general population. Simply expanding the sample size of a general household survey (e.g., the CPS) is prohibitively expensive because the "yield" (the number of sample cases of interest) is so low—about 5 cases for each 100 people added to the sample in this instance.

The NSF Approach

Historically, NSF has used the decennial census long-form sample as a screening device for obtaining adequate samples of scientists and engineers. NSF sponsored surveys drawn from census respondents following the 1960, 1970, 1980, and 1990 censuses. Individuals identified as

---

[1] The sample size for scientists and engineers in the CPS could be expanded by combining monthly samples for a year. However, the increase in effective sample size would be much less than 12 times the monthly sample size because of the rotation group design used in the CPS, whereby household addresses are included in the survey for 4 months, dropped from the survey for 8 months, and then included in the survey for another 4 months. Also, the richest CPS data for analysis of scientists and engineers are available only for March, when the annual income supplement is administered.

scientists and engineers in the 1972, 1982, and 1993 postcensal surveys were resurveyed over the decade at 2-year intervals. Also, surveys were conducted regularly of new entrants to the field, specifically, new bachelor's and master's degree recipients in science and engineering disciplines identified by institutions of higher education. In addition, NSF funded continuing surveys of Ph.D.-level scientists and engineers. This approach has several problems, however, of which two are of particular significance.

(1) First, the census long-form questionnaire is not a particularly efficient screener for scientists and engineers. It is possible to use the data on current occupation to oversample people who report they worked as a scientist or engineer and to use the data on level of education attained to limit the sample to college graduates. However, the census does not ask about degree field, and NSF has never been able to get such a question included in the census. Hence, it is not possible with the census data to oversample people with science and engineering degrees. Given user interest not only in people who work as scientists and engineers, but also in those who are trained in science and engineering fields, the deficiencies of the census questionnaire present significant challenges for a cost-effective sample design.

One approach is to draw a large sample not only of people reporting science and engineering occupations, but also of other college graduates. This approach quickly becomes expensive because the stratum of other college graduates includes many cases who are not of interest. Another approach is to sample other college graduates at a low rate relative to working scientists and engineers, which permits a smaller overall sample size. However, if the differences in the sampling rates are too great, there will be substantial increases in the standard errors of estimates that are based—as will usually be the case—on both strata (working scientists and engineers and other college graduates). Differences in sampling rates of 100 to 1 were used in the NSF 1982 postcensal survey, and the results were disastrous for the quality of the estimates.[i]

The CNSTAT Panel to Study the NSF Scientific and Technical Personnel Data System recommended that the differences in sampling rates be no more than 4 to 1; as a compromise, the design of the 1993 postcensal survey (the National Survey of College Graduates) has sampling rates that vary by no more than 8 to 1. To reduce the differences between sampling rates while maintaining the precision of estimates of scientists and engineers, the overall size of the 1993 NSCG was increased to 215,000 initial sample cases from 138,000 cases in the 1982 postcensal survey. Because of its large sample size, the postcensal survey has to be conducted by mail to be affordable, which, in turn, limits the kind and amount of detail that can be ascertained.

(2) Second, the approach of using the decennial census as a screener to identify the stock of scientists and engineers to follow up over the decade together with new graduates in science and engineering fields means that the NSF data system cannot readily identify some population groups of interest. There is no cost-effective way to add these groups to the sample.

One such group is people who, during the decade, move into science and engineering jobs from non-science and engineering backgrounds. The computer science field is one example in which a significant number of people who work in that field were not trained in a science or engineering discipline. This kind of movement across the boundaries between science and engineering and other disciplines cannot be identified in the NSF system until the next postcensal survey.

Another population group of interest that the NSF approach misses is scientists and engineers who enter the United States during the decade who were not trained in the United States. While there are other sources of data about immigrants, they are problematic in both level of detail and quality. Again, new immigrants cannot be captured in the NSF data system until the next postcensal survey.

A Better Approach for the Future?

Two new surveys offer the possibility that NSF could develop a cost-effective science and engineering personnel data system that responds more fully to user needs. One survey, which began in 1994, is the National Immunization Survey (NIS), sponsored by the Centers for Disease Control in the Department of Health and Human Services. The second survey, which is scheduled to become fully operational in 2003, is the American Community Survey (ACS) that is being developed by the Census Bureau.

The NIS is a random digit dialing (RDD) survey in which about 3 million telephone numbers per year are called to identify families with young children, who are asked to respond to questions about immunization. It is possible that NSF could pay to have questions added to the NIS screener to identify people working or trained as scientists and engineers, who would be asked to respond to an NSF-designed questionnaire. RDD surveys miss that portion of the population without telephones, but this lack should not be a problem for the science and engineering population. To minimize respondent burden for families with both a scientist or engineer and young children, the sample design could eliminate overlap by assigning a portion of such families the immunization questionnaire and a portion the NSF questionnaire.

To maintain the desired sample sizes for both NSF and CDC, NSF may need to provide funding not only for added questions, but also for added screening interviews. Even so, the costs of piggybacking on the NIS would likely compare favorably with the costs of the current system. Moreover, such an approach should afford the opportunity to include such population groups as immigrants and, perhaps, to have a more richly detailed questionnaire than in the current system.

The ACS is designed as a mailout-mailback survey, with telephone and personal follow-up, in which 250,000 households are to be surveyed each month, for a total of 3 million households per year (about 7-8 million people), with no overlap in the samples across months. The survey is currently being tested in pilot sites and is planned to become fully operational beginning in 2003. If implementation of the ACS proceeds as planned, it will likely replace the census long-form questionnaire in the year 2010. The ACS questionnaire will include the basic long-form data, including educational level and occupation. With NSF support, it could be possible to add a question on degree field, which would permit using the responses to identify both working and trained scientists and engineers for a follow-up mail survey. If respondents provide telephone numbers, it could be possible to conduct the NSF follow-up survey by phone.

With either the NIS or the ACS, NSF would need to decide how frequently over the course of a decade to use one or the other survey as a screener. One approach would be to begin the decade by using the NIS or ACS to obtain a large sample of scientists and engineers who would then be followed up at regular intervals. At these same intervals, the NIS or ACS could be used to identify new entrants (e.g., immigrants, new degree recipients) to add to the longitudinal sample. Alternatively, the NIS or ACS could be the source of regular cross-sectional surveys of scientists and engineers. Also, one or the other survey could be used on occasion to identify subsamples of scientists and engineers to receive questionnaires on special topics.

Box D1 provides approximate sample sizes from the SESTAT system, the March CPS, and the ACS for various size groups of scientists and engineers to illustrate the gains from use of the ACS. (The gains would presumably be similar for the NIS.) The estimates for the ACS are derived by assuming that every case of interest receives the NSF questionnaire; in fact, it would be possible to subsample to reduce costs. The sample sizes shown for the CPS and ACS are approximate and do not take account of sample design features that could reduce the effective sample size.

No information on costs is available at present, but it would be worthwhile for NSF to investigate the costs and benefits of using the ACS or the NIS versus the current approach in the future.

Box D-1. Illustrative Approximate Sample Sizes for Estimates of Scientists and Engineers: NSF SESTAT, March CPS, ACS

Total scientists and engineers--12 million:

Sample size SESTAT:	100,000
Sample size CPS (1 mo.):	6,000
Sample size ACS (12 mos):	300,000

500,000 group:

Sample size SESTAT:	4,200
Sample size CPS:	250
Sample size ACS:	12,000

50,000 group:

Sample size SESTAT:	420
Sample size CPS:	25
Sample size ACS:	1,200

Note: ACS sample sizes have been reduced by about 25 percent to allow for nonresponse.

---

[i] Panel to Study the NSF Scientific and Technical Personnel Data System, *Surveying the Nation's Scientists and Engineers*, (Washington, DC: National Academy Press, 1989).

Appendix E

Biographical Sketches of Committee Members

Biographical Sketches of Committee Members

Janice Madden (Chair) is Vice Provost, Graduate Education, University of Pennsylvania. Dr. Madden, who has been on the faculty at the University of Pennsylvania since 1972, serves as the Robert C. Daniels Term Professor of Urban Studies, Regional Science, Sociology, and Real Estate, and as a Professor in the Department of Real Estate at the Wharton School. She is also a Research Associate at the University of Pennsylvania's Population Studies Center and has previously served as Director of the Alice Paul Research Center and the Women's Studies Program at the University. Dr. Madden has been on the Board of Directors of, and a consultant with, Econsult Corporation of Philadelphia since 1980. Her clients have included the U.S. Army Family Research Program, the U.S. Equal Employment Opportunity, and the U.S. Department of Justice. Her recent publications include "Work, Wages, and Poverty: Income Distribution in Post-Industrial Philadelphia,' with William Stull, (1991) and "Rising Incomes and Earnings Inequality: U.S. Metropolitan Areas in the 1980s" (forthcoming). She serves on the editorial board for *Women and Work* and is the U.S. editor for *Urban Studies*. Dr. Madden has just completed a term as President of the Association of Graduate Schools (AGS) and as a member of the American Association of Universities' (AAU's) Committee on Graduate Education. She is currently a member of the Board of Directors of the Council of Graduate Schools, is a member of the Graduate Record Examination Board, and serves on the Steering Committee for the AAU/AGS Project for Research on Doctoral Education. Dr. Madden has previously served on the NRC's Committee on Vocational Education and Economic Development in Depressed Areas. Her honors include, most recently, the Academic Excellence Award of the Trustees' Council of Penn Women (1997). She holds a Ph.D. (1972) in economics from Duke University and a B.A. (1969) in economics from the University of Denver.

Paul R. Biemer is Chief Scientist and Director of the Survey Methods Research Program at Research Triangle Institute, Inc., in Research Triangle Park, North Carolina. He also serves as an instructor in both the Survey Research Center, Annual Summer Institute, University of Michigan and the Joint University of Maryland-University of Michigan Program in Survey Methodology. Before joining RTI in 1991, he was head of the Department of Experimental Statistics and director of the University Statistics Center at New Mexico State University. Prior to that, he was assistant chief of the Statistical Research Division at the Bureau of the Census. Dr. Biemer has more than 18 years of postdoctoral experience in survey methods and statistics, including survey methodology, nonsampling and measurement error evaluation and analysis, survey design and estimation, experimental design and analysis, and contract research management. He specializes in the design and analysis of studies to evaluate alternative survey designs and has published widely in these areas. Dr. Biemer serves as associate editor for the *Journal of Official Statistics* and is a reviewer for numerous other journals. He is a Fellow of the American Statistical Association. He has previously chaired the Survey Research Methods Section of the American Statistical Association and has served on the NRC's Panel to Evaluate the Survey of Income and Program Participation. Dr. Biemer received the Bronze Medal from the Bureau of the Census for superior federal service in 1985 and the H.O. Hartley Award for outstanding contributions to the statistical profession 1990. He received a B.S. (1972) in mathematics and a Ph.D. (1978) in statistics from Texas A&M University.

Bronwyn Hall is Associate Professor, Department of Economics at the University of California, Berkeley where she has been on the faculty since 1987. She has several simultaneous appointments: since 1996, she has also been serving as Temporary Professor of Economics and Fellow of Nuffield College, Oxford University; since 1995 she has been International Research Associate, Institute for Fiscal Studies, London, England; and since 1988 she has been a Research Fellow or Associate, National Bureau of Economic Research, and a member of its Programs on Productivity and Technical Change. Since 1977 she has also been founder and owner, TSP International, a computer software firm that maintains, develops, and distributes the TSP econometrics package. Dr. Hall's research interests focus on the economics of research and development and innovation. She is Associate Editor, *Economics of Innovation and New Technology*, and a member of the advisory board for *International Finance*. She has also been a member of the editorial board for *Economics of Innovation and New Technology*. Dr. Hall has served on the Census Advisory Committee of the American Economic Association; the Advisory Committee to the Panel Study of Entrepreneurial Dynamics, Institute for Social Research, University of Michigan; the Advisory Committee of Economists to the Inter-University Consortium for Political and Social Research; and the Data Base Review Committee of the Small Business Administration, 1984-1985. She has served on two NRC panels, the Steering Committee for a Workshop on Industrial Science and Technology Indicators and the Steering Committee on Projections of Scientists and Engineers. She has also testified before the House Subcommittee on Science, Space, and Technology. Dr. Hall holds a B.A. in physics (1966) from Wellesley College and a Ph.D. in economics (1988) from Stanford University.

T.R. Lakshmanan is Professor, Department of Geography, and Executive Director, Center for Energy and Environmental Studies, at Boston University. Dr. Lakshmanan held these positions from 1978 and 1979, respectively, until 1994, and resumed them in January 1998. From 1994 to 1998, Dr. Lakshmanan served as director of the Bureau of Transportation Statistics at the U.S. Department of Transportation. Dr. Lakshmanan has also been a Visiting Scholar, International Institute for Applied Systems Analysis; Fellow, Clare Hall, Cambridge University; Visiting Scholar, Massachusetts Institute of Technology; and Visiting Scholar, The Institute for Futures Studies, Stockholm. He is the author of numerous books, articles, and papers. He has authored and edited *Systems and Models for Energy and Environmental Analysis*; *Spatial, Environmental and Resource Policy in Developing Countries*; *Rural Industrialization in Regional Development in the Third World*; *Large-Scale Energy Projects: Assessment of Regional Consequences*; and *Economic Faces of the Building Sector*. His recent articles include "Full Benefits and Costs of Transportation: Review and Prospects," "Technical Change in Transportation: Social and Institutional Issues," and "The Changing Context of Transportation Modeling: Implications of the New Economy, Intermodalism, and the Drive for Environmental Quality." Dr. Lakshmanan was editor, *Annals of Regional Science*, from 1988 to 1994. He served as Chairman of the Working Group on Energy Resources and Development of the International Geographic Union from 1980 to 1988 and was Vice President of the International Regional Science Association from 1981 to 1983. He has served on the Executive Committee of NRC's Transportation Research Board and on the Panel on Technologies for Affordable Housing. He is currently serving on the NRC's Committee on Geography. In 1985 he was elected a Life Member of Clare Hall College, Cambridge University and in 1989 was awarded the Anderson Medal of the American Association of Geographers. Dr. Lakshmanan holds a Ph.D. (1965) from Ohio State University and an M.A. (1953) and B.Sc. (1952) from the University of Madras.

Eduardo R. Macagno is Associate Vice President for Research and Graduate Education and Dean of the Graduate School of Arts and Sciences at Columbia University. Dr. Macagno holds a faculty position as professor of biological sciences in the department of Biological Sciences at Columbia where he has been since 1973. Dr. Macagno also served as an instructor in neural systems and behavior at the Marine Biological Laboratory for fifteen years. He has sponsored nineteen graduate students and eight postdoctoral fellows at Columbia. Dr. Macagno's laboratory studies the elucidation of the cellular and molecular mechanisms underlying features of the developing nervous system. He currently holds grants from the National Science Foundation to study "Receptor Phosphatases and Control of Neurite Growth" and from the National Institutes of Health to study "Cell Interactions and the Genesis of Neuronal Arbors." He has numerous publications in physics and neurobiology. His edited books include P.C. Letourneau, S.B. Kater, and E.R. Macagno, eds., *The Nerve Growth Cone* (1991) and M. Shankland and E.R. Macagno, eds., *Determinants of Cell Fate* (1992). Dr. Macagno is co-editor of the *Journal of Neurobiology*. He was a member of the Cell and Molecular Basis of Disease Study Section of the National Institutes of Health from 1991 to 1996 and chaired the Section from 1994 to 1996. He served on the American Association of Universities' Committee on the Future of Graduate Education from 1996 to 1998. He was selected Fellow of the American Association for the Advancement of Science in 1992. Dr. Macagno holds a Ph.D. (1968) in physics from Columbia University and a B.A. (1963) in physics from the University of Iowa.

Robert H. McGuckin is Director of Economic Research at the Conference Board with which he has served since August 1996. In addition to his responsibilities for economic research at the Conference Board, Dr. McGuckin supervises the Business Cycle Indicators program and the Consumer Research Center. Prior to joining the Conference Board, Dr. McGuckin was chief of the Center for Economic Studies (CES) at the U.S. Bureau of the Census where he guided development of the Longitudinal Research Database (LRD) and a broad research program in both statistics and economics. During his tenure, CES became a world leader in the development of microdata approaches to economic theory and policy. Dr. McGuckin held several positions with the Antitrust Division of the U.S. Department of Justice during the years 1974-86, including Assistant Director of the Economic Policy office and Director of Research for the Economic Analysis Group. While at the Department of Justice, Dr. McGuckin was named the Victor H. Kramer Fellow at the University of Chicago School of Law for the 1978-1979 academic year. Before entering government service, Dr. McGuckin was an Assistant Professor of Economics at the University of California at Santa Barbara from 1970 to 1976. Dr. McGuckin is a specialist in industrial organization, productivity, antitrust, and statistics, and has published numerous articles on economic and statistical topics in refereed professional journals. His recent work has focused on mergers and acquisitions, adoption of advanced computer technologies by business, and business organizations, and he has also written on a wide range of topics, from the effects of Chinese economic reforms on productivity growth to issues of the effect of aggregation on economic studies. Dr. McGuckin currently serves as a director of the Center for the Study of Contracts and Structure of Industry at the University of Pittsburgh's Katz School of Business. He received his Ph.D. in economics from the State University of New York at Buffalo in 1970, and his B.A. in mathematics from Ithaca College in 1965.

John McTague recently retired as Vice President, Technical Affairs, Ford Motor Company. Appointed to the position in 1990, he directed the operations of Ford Research Laboratories, Environmental and Safety Engineering, New Generation Vehicle Programs, and Corporate Technology Planning. He previously served as Ford's Vice President, Research. Prior to joining Ford in 1986, Dr. McTague served as Acting Science Advisor to the President of the United

States. He has also held positions as director of the National Synchotron Light Source at Brookhaven National Laboratory in New York, as adjunct professor of chemistry at Columbia University, as a faculty member in chemistry at the University of California, Los Angeles, and as a member of the technical staff of the North American Rockwell Science Center. Dr. McTague was appointed by President Bush to his Council of Advisors on Science and Technology (PCAST) in February 1990. He is a former member of the Advisory Board for the Directorate on Social, Behavioral and Economic Sciences at the National Science Foundation. Dr. McTague is a member of the National Academy of Engineering and he has served on a number of NRC Committees, including most recently the Steering Committee on Projections of Scientists and Engineers. He is a member of the Secretary of Energy's Advisory Board and is Chairman of the Board of Overseers of Fermilab. He serves on the boards of directors of the National Action Council for Minorities in Engineering, the State of Michigan Strategic Fund, Michigan Technologies, Inc., and Raychem Corporation. He holds a B.S. (1960) and a Ph.D. (1965) in physical chemistry from Brown University.

David Mowery is Milton W. Terrill Professor of Business at the Walter A. Haas School of Business at the University of California, Berkeley, and Director of the Haas School's Ph.D. Program. Prior to joining the faculty at Berkeley, Dr. Mowery taught at Carnegie-Mellon University, served as Study Director for the NRC Panel on Technology and Employment, and served in the Office of the United States Trade Representative as a Council on Foreign Relations International Affairs Fellow. His research deals with the economics of technological innovation and with the effects of public policies on innovation. Dr. Mowery has published numerous academic papers and has written or edited a number of books, including *The International Computer Software Industry: A Comparative Study of Industry Evolution and Structure*; *Paths of Innovation: Technological Change in 20th-Century America*; *The Sources of Industrial Leadership*; *Science and Technology Policy in Interdependent Economies*; *Technology and the Pursuit of Economic Growth*; *Alliance Politics and Economics: Multinational Joint Ventures in Commercial Aircraft*; *Technology and Employment: Innovation and Growth in the U.S. Economy*; *The Impact of Technological Change on Employment and Economic Growth*; *Technology and the Wealth of Nations*; and *International Collaborative Ventures in U.S. Manufacturing*. Dr. Mowery has served on a number of National Research Council panels, including those on the Competitive Status of the U.S. Civil Aviation Industry, the Causes and Consequences of the Internationalization of U.S. Manufacturing, the Federal Role in Civilian Technology Development, U.S. Strategies for the Children's Vaccine Initiative, and Applications of Biotechnology to Contraceptive Research and Development. He has also testified before congressional committees and served as an adviser for the Organization for Economic Cooperation and Development, various federal agencies and industrial firms. His academic awards include the Raymond Vernon Prize from the Association for Public Policy Analysis and Management, the Economic History Association's Fritz Redlich Prize, the Business History Review's Newcomen Prize, and the Cheit Outstanding Teaching Award. He received his undergraduate and Ph.D. degrees in economics from Stanford University and was a postdoctoral fellow at the Harvard Business School.

Julie Norris is Director, Sponsored Programs, Massachusetts Institute of Technology, a position she assumed after a long career at the University of Houston where she was Assistant Vice President and Director of Sponsored Programs. Her responsibilities at MIT include management of both pre- and post-award activities in the area of sponsored programs, including responsibility for the preparation and negotiation of the institute's indirect cost proposal and other cost analysis activities. Ms. Norris is a member of the Council on Governmental Relations (COGR). She served on COGR's board from 1982-1988 and was chair of the Grant and Contract Policy Committee in 1987-88. She was re-appointed to the board in 1992 for another six year term, served as chair of the Grant and Contract Policy Committee from 1992-1994, and as Chairman of the Board, 1994-1996. Ms. Norris served as chair of COGR's Costing Policies Committee in 1996-1997 and for 1997-1998 serves as chair of the Research Administration and Compliance Committee. She is also a member of the National Council of University Research Administrators (NCURA) and has served that national organization as treasurer, vice president, and president. She has also been a consultant to the National Science Foundation on its Research Facilities and Expenditures studies. Ms. Norris served on the research team for the study entitled "Financing and Managing University Research Equipment" which was produced by the American Association of Universities (AAU), the National Association of State Universities and Land Grant Colleges (NASULGC), and COGR. She is the primary contributor to the COGR document "Managing Externally Funded Programs at Colleges and Universities" and one of the authors of NCURA's "Regulation and Compliance Handbook." She is the author of Volume I of NCURA's "Fundamentals of Sponsored Projects Administration" and is currently working on Volume II. In addition she is the author of the sponsored programs chapter in NACUBO's "College and University Business Administration" and one of the authors of AIS's "Managing Federal Grants." Ms. Norris was the first recipient of NCURA's award for Outstanding Contributions to Research Administration. She holds an M.A. (1966) in history from the University of Houston, and a B.A. (1958) in history from Rice University.

Paula E. Stephan is Professor of Economics and Associate Dean of Policy Studies at Georgia State University where she has been on the faculty since 1971. A labor economist by training, her recent research focuses on issues in science and technology. She is interested in both the careers of scientists and engineers and the process by which knowledge moves across institutional boundaries in the economy. Her publications include "The Economics of Science," Journal of Economics Literature (September 1996); "Company Scientist Locational Links: The Case of Biotechnology," (with D. Audretsch), *American Economic Review* (June 1996); and *Striking the Mother Lode in Science: The Importance of Age, Place, and Time* (with S. Levin), Oxford University Press, 1992. Dr. Stephan has received funding for research from the North Atlantic Treaty Organization, the National Science Foundation, the Alfred P. Sloan Foundation, the U.S. Department of Labor, the Exxon Education Foundation, and the Andrew Mellon Foundation. Dr. Stephan has recently served on several NRC Committees, including the Committee on Dimensions, Causes, and Implications of Trends in the Early Career Events for Life Scientists; Committee on Methods of Forecasting Demand and Supply of Doctoral Scientists and Engineers; and the Committee on NRC Research Associates Career Outcomes. She holds a B.A. from Grinnell College and an M.A. and Ph.D. in economics from the University of Michigan.